HANDBOOK OF COAL ANALYSIS

CHEMICAL ANALYSIS

A SERIES OF MONOGRAPHS ON ANALYTICAL CHEMISTRY
AND ITS APPLICATIONS

Series Editor
J. D. WINEFORDNER

VOLUME 166

HANDBOOK OF COAL ANALYSIS

James G. Speight

A JOHN WILEY & SONS, INC., PUBLICATION

Published by John Wiley & Sons, Inc., Hoboken, New Jersey.
Published simultaneously in Canada.

For general information on our other products and services please contact our Customer Care Department within the U.S. at 877-762-2974, outside the U.S. at 317-572-3993 or fax 317-572-4002.

Wiley also publishes its books in a variety of electronic formats. Some content that appears in print, however, may not be available in electronic format.

Library of Congress Cataloging-in-Publication Data:

Speight, J. G.
Handbook of coal analysis / James G. Speight.
p. cm.
Includes bibliographical references and index.
ISBN 0-471-52273-2 (cloth)
1. Coal—Analysis—Handbooks, manuals, etc. I. Title.

TP325.S7145 2005
662.6'22—dc22

2005040803

Printed in the United States of America.

10 9 8 7 6 5 4 3 2 1

CONTENTS

PREFACE

Coal is an extremely complex material and exhibits a wide range of physical properties. The rapidly expanding use of coal in the twentieth century made it necessary to devise acceptable methods for coal analysis with the goal of correlating composition and properties with behavior. It is only by assiduously using careful analyses of coal that the various aspects of coal usage can be achieved in an environmentally acceptable manner. As a part of the multifaceted program of coal evaluation, new methods are continually being developed and the already accepted methods may need regular modification to increase the accuracy of the technique as well as the precision of the results. Furthermore, proper interpretation of the data resulting from the analysis of coal requires an understanding of the significance of the analytical data.

This book deals with the various aspects of coal analysis and provides a detailed explanation of the necessary standard tests and procedures that are applicable to coal in order to help define coal behavior relative to usage and environmental issues. The first items that the book covers (after nomenclature and terminology) are related to sampling, accuracy of analysis, and precision of analysis. These important aspects are necessary to provide reproducibility and repeatability of the analytical data derived from the various test methods. The book then goes on to provide coverage of the analysis of coal by various test methods, as well as the application and interpretation of the data to provide the reader with an understanding the quality and performance of coal. A glossary of terms that will be useful to the reader is also included. Each term is defined in a language that will convey the meaning to the reader in a clear and understandable way.

Sources of information that have been used include (1) the *Annual Book of ASTM Standards*, (2) the British Standards Institution, (3) the International Organization for Standardization, (4) older books, (5) collections of individual articles from symposia, and (6) chapters in general coverage books. This will be the first book that provides not only a detailed description of the tests but also the outcome of the tests and the meaning of the data. However, the actual mechanics of performing a test method are not included; such information is available from the various standards organization.

Although the focus of the book is on the relevant ASTM (American Society for Testing and Materials) test methods with the numbers given, where possible the corresponding ISO (International Organization for Standardization) and BS (British Standards Institution) test method numbers are also presented. As an aside, the ASTM may have withdrawn some of the test methods noted herein, but the test method is still referenced because of its continued use, for whatever reason, in many analytical laboratories.

If this book helps toward a better understanding of the criteria for determining the properties of coal, leading to an understanding of coal behavior, it will have served its purpose.

James G. Speight

Laramie, Wyoming
August 2004

1 Coal Analysis

Coal is an organic sedimentary rock that contains varying amounts of carbon, hydrogen, nitrogen, oxygen, and sulfur as well as trace amounts of other elements, including mineral matter (van Krevelen, 1961; Gluskoter, 1975; Speight, 1994; ASTM D-121).

The name *coal* is thought to be derived from the Old English *col*, which was a type of charcoal used at the time. Coal is also referred to in some areas, as *sea coal* because it is occasionally found washed up on beaches, especially in northeastern England. Generally, coal was not mined to any large extent during the early Middle Ages (prior to A.D. 1000) but there are written records of coal being mined after that date. However, the use of coal expanded rapidly, throughout the nineteenth and early twentieth centuries. This increased popularity has made it necessary to devise acceptable methods for coal analysis, with the goal of correlating fuel composition and properties with behavior (Montgomery, 1978; Vorres, 1993; Speight, 1994).

Coal is a solid, brittle, combustible, carbonaceous rock formed by the decomposition and alteration of vegetation by compaction, temperature, and pressure. It varies in color from brown to black and is usually stratified. The source of the vegetation is often moss and other low plant forms, but some coals contain significant amounts of materials that originated from woody precursors.

The plant precursors that eventually formed coal were compacted, hardened, chemically altered, and metamorphosed by heat and pressure over geologic time. It is suspected that coal was formed from prehistoric plants that grew in swamp ecosystems. When such plants died, their biomass was deposited in anaerobic, aquatic environments where low oxygen levels prevented their reduction (rotting and release of carbon dioxide). Successive generations of this type of plant growth and death formed deep deposits of unoxidized organic matter that were subsequently covered by sediments and compacted into carboniferous deposits such as peat or bituminous or anthracite coal. Evidence of the types of plants that contributed to carboniferous deposits can occasionally be found in the shale and sandstone sediments that overlie coal deposits.

Coal deposits, usually called *beds* or *seams*, can range from fractions of an inch to hundreds of feet in thickness. Coals are found in all geologic periods from Silurian through Quaternary, but the earliest commercially important coals are found in rocks of Mississippian age (Carboniferous in Europe). Coals generally

Handbook of Coal Analysis, by James G. Speight
ISBN 0-471-52273-2

formed either in basins in fluvial environments or in basins open to marine incursions. Coal is found on every continent, and world coal reserves exceed 1 trillion tons. However, the largest reserves are found in the United States, the former Soviet Union, and China. The United States and former Soviet Union each have about 23% of the world's reserves, and China has about 11%.

Coal consists of more than 50% by weight and more than 70% by volume of carbonaceous material (including inherent moisture). It is used primarily as a solid fuel to produce heat by burning, which produces carbon dioxide, a greenhouse gas, along with sulfur dioxide. This produces sulfuric acid, which is responsible for the formation of sulfate aerosol and acid rain. Coal contains many trace elements, including arsenic and mercury, which are dangerous if released into the environment. Coal also contains low levels of uranium, thorium, and other naturally occurring radioactive isotopes, whose release into the environment may lead to radioactive contamination. Although these substances are trace impurities, a great deal of coal is burned, releasing significant amounts of these substances.

When coal is used in electricity generation, the heat is used to create steam, which is then used to power turbine generators. Approximately 40% of Earth's current electricity production is powered by coal, and the total known deposits recoverable by current technologies are sufficient for at least 300 years of use. Modern coal power plants utilize a variety of techniques to limit the harmfulness of their waste products and to improve the efficiency of burning, although these techniques are not widely implemented in some countries, as they add to the capital cost of the power plant.

Coal exists, or is classified, as various types, and each type has distinctly different properties from the other types. *Anthracite*, the highest rank of coal, is used primarily for residential and commercial space heating. It is hard, brittle, and black lustrous coal, often referred to as *hard coal*, containing a high percentage of fixed carbon and a low percentage of volatile matter. The moisture content of fresh-mined anthracite generally is less than 15%. The heat content of anthracite ranges from 22 to 28 million Btu/ton on a moist, mineral-matter-free basis.

Bituminous coal is a dense coal, usually black, sometimes dark brown, often with well-defined bands of bright and dull material, used primarily as fuel in steam-electric power generation, with substantial quantities also used for heat and power applications in manufacturing and to make coke. The moisture content of bituminous coal is usually less than 20% by weight. The heat content of bituminous coal ranges from 21 to 30 million Btu/ton on a moist, mineral-matter-free basis.

Subbituminous coal is coal whose properties range from those of lignite to those of bituminous coal, used primarily as fuel for steam-electric power generation. It may be dull, dark brown to black, and soft and crumbly at the lower end of the range, to bright, black, hard, and relatively strong at the upper end. Subbituminous coal contains 20 to 30% inherent moisture by weight. The heat content of subbituminous coal ranges from 17 to 24 million Btu per ton on a moist, mineral-matter-free basis.

Lignite is the lowest rank of coal, often referred to as *brown coal*, used almost exclusively as fuel for steam-electric power generation. It is brownish black and has a high inherent moisture content, sometimes as high as 45%. The heat content of lignite ranges from 9 to 17 million Btu/ton on a moist, mineral-matter-free basis.

1.1 ANALYSIS CONSIDERATIONS

The data obtained from coal analyses (Table 1.1) establish the price of the coal by allocation of production costs as well as to control mining and cleaning operations and to determine plant efficiency. However, the limitations of the analytical methods must be recognized (Rees, 1966). In commercial operations, the price of coal not only reflects the quantity of coal but also invariably reflects the relationship of a desirable property or even a combination of properties to performance of coal under service conditions (Vorres, 1993).

Measurements of the desired property or properties (usually grouped together under the general title *specifications*) are expressed as numerical values; therefore, the accuracy of these measurements is of the utmost importance. The measurements need to be sufficiently accurate so as to preclude negative scientific or economic consequences. In other words, the data resulting from the test methods used must fall within the recognized limits of error of the experimental procedure so that the numerical data can be taken as fixed absolute values and

TABLE 1.1 Sampling and Analytical Methods Used for Coal Evaluation

Test/Property	Results/Comments
Sample information	
Sample history	Sampling date, sample type, sample origin (mine, location)
Sampling protocols	Assurance that sample represents gross consignment
Chemical properties	
Proximate analysis	Determination of the "approximate" overall composition (i.e., moisture, volatile matter, ash, and fixed carbon content)
Ultimate analysis	Absolute measurement of the elemental composition (i.e., carbon, hydrogen, sulfur, nitrogen, and oxygen content)
Sulfur forms	Chemically bonded sulfur: organic, sulfide, or sulfate
Ash properties	
Elemental analysis	Major elements
Mineralogical analysis	Analysis of the mineral content
Trace element analysis	Analysis of trace elements; some enrichment in ash
Ash fusibility	Qualitative observation of temperature at which ash passes through defined stages of fusing and flow

Source: Smith and Smoot (1990).

not as approximations. Indeed, the application of statistical analysis to such test methods must be treated with extreme caution. Such analysis must be based on valid assumptions and not be subject to a claim of mathematical manipulation to achieve the *required result*. In other words, there is a requirement that reliable *standard* test methods be applied to coal analysis.

There are many problems associated with the analysis of coal (Lowry, 1963; Karr, 1978) not the least of which is its heterogeneous nature. Other problems include the tendency of coal to gain or lose moisture and to undergo oxidation when exposed to the atmosphere. In addition, the large number of tests and analyses required to characterize coal adequately also raise issues.

Many of the test methods applied to coal analysis are empirical in nature, and strict adherence to the procedural guidelines is necessary to obtain repeatable and reproducible results. The type of analysis normally requested by the coal industry may be a proximate analysis (moisture, ash, volatile matter, and fixed carbon) or an ultimate analysis (carbon, hydrogen, sulfur, nitrogen, oxygen, and ash).

By definition, a *standard* is defined as a document, established by consensus and approved by a recognized body, that provides, for common and repeated use, rules, guidelines, or characteristics for activities or their results. Many industry bodies and trade associations require a product (e.g., coal) to conform to a standard or directive before it can be offered for sale. In fact, the use of standards is becoming more and more of a prerequisite to worldwide trade. Above all, any business, large or small, can benefit from the conformity and integrity that standards assure.

As a result, the formation of various national standards associations has led to the development of methods for coal evaluation. For example, the American Society for Testing and Materials (ASTM) has carried out uninterrupted work in this field for many decades, and investigations on the development of the standardization of methods for coal evaluation has occurred in all the major coal-producing countries (Montgomery, 1978; Patrick and Wilkinson, 1978). There are in addition to the ASTM, organizations for methods development and standardization that operate on a national level; examples are the International Organization for Standardization (ISO) and the British Standards Institution (BS), which covers the analysis of coal under one standard number (BS 1016) (Table 1.2).

Furthermore, the increased trade between various coal-producing countries that followed World War II meant that cross-referencing of already accepted standards was a necessity, and the mandate for such work fell to the ISO, located in Geneva, Switzerland; membership in this organization is allocated to participating (and observer) countries. Moreover, as a part of the multifaceted program of coal evaluation, new methods are continually being developed and the methods already accepted may need regular modification to increase the accuracy of the technique as well as the precision of the results.

It is also appropriate that in any discussion of the particular methods used to evaluate coal for coal products, reference should be made to the relevant test. Accordingly, where possible, the necessary test numbers (ASTM) have been included as well as those, where appropriate, of the BS and the ISO.

TABLE 1.2 British Standard 1016: ***Methods for Analysis and Testing of Coal and Coke***

Section	Topics
BS 1016-1	Total moisture of coal
BS 1016-6	Ultimate analysis of coal
BS 1016-7	Ultimate analysis of coke
BS 1016-8	Chlorine in coal and coke
BS 1016-9	Phosphorus in coal and coke
BS 1016-10	Arsenic in coal and coke
BS 1016-14	Analysis of coal ash and coke ash
BS 1016-21	Determination of moisture-holding capacity of hard coal
BS 1016-100	General introduction and methods for reporting results
BS 1016-102	Determination of total moisture of coke
BS 1016-104.1	Proximate analysis, determination of moisture content of the general analysis test sample
BS 1016-104.2	Proximate analysis, determination of moisture content of the general analysis sample of coke
BS 1016-104.3	Proximate analysis, determination of volatile matter content
BS 1016-104.4	Proximate analysis, determination of ash content
BS 1016-105	Determination of gross calorific value
BS 1016-106.1.1	Ultimate analysis of coal and coke, determination of carbon and hydrogen content, high temperature combustion method
BS 1016-106.1.2	Liebig method
BS 1016-106.2	Ultimate analysis of coal and coke, determination of nitrogen content
BS 1016-106.4.1	Ultimate analysis of coal and coke, determination of total sulfur content, Eschka method
BS 1016-106.4.2	Ultimate analysis of coal and coke, determination of total sulfur content, high temperature combustion method

Source: BS (2003).

A complete discussion of the large number of tests that are used for the evaluation of coal (and coal products) would fill several volumes (see, e.g., Ode, 1963; Karr, 1978, 1979; Montgomery, 1978; Zimmerman, 1979; Gluskoter et al., 1981; Smith and Smoot, 1990), and such detailed treatment is not the goal of this book. The focus is on a description, with some degree of detail, of the test methods in common use, as well as a critique of various procedures that are not obvious from the official descriptions of test methods and a description of pitfalls that can occur during application of a test method for coal analysis.

Quite often, a variation of a proximate analysis or an ultimate analysis is requested, together with one or more of the miscellaneous analyses or tests discussed in this chapter. Restrictions that have been placed on the coal used in coal-fired power plants and other coal-burning facilities have created a need for more coal analyses as well as a need for more accurate and faster methods of analysis. This trend will continue, and more testing will be required with increased use of coal in liquefaction and gasification plants.

1.2 ACCURACY AND PRECISION

In any form of analysis, accuracy and precision are required; otherwise, the analytical data are suspect and cannot be used with any degree of certainly. This is especially true of analytical data used for commercial operations where the material is sold on the basis of *purity*. Being a complex material, one may wonder about the purity of coal, but in this sense the term *purity* refers to the occurrence (or lack thereof) of foreign constituents within the organic coal matrix. Such foreign constituents (impurities) are water, pyrite, and mineral matter. Therefore, at this point, it is advisable to note the differences inherent in the terms *accuracy* and *precision*.

The word *accuracy* is used to indicate the reliability of a measurement or an observation, but it is, more specifically, a measure of the closeness of agreement between an experimental result and the true value. Thus, the accuracy of a test method is the degree of agreement of individual test results with an accepted reference value.

On the other hand, *precision* is a measure of the degree to which replicate data and/or measurements conform to each other, the degree of agreement among individual test results obtained under prescribed similar conditions. Hence, it is possible that data can be very precise without necessarily being correct or accurate. These terms will be found throughout any book devoted to a description of standard methods of analysis and/or testing, and have sometimes been used (incorrectly) interchangeably. Precision is commonly expressed inversely by the imprecision of results in terms of their standard deviation or their variance. Precision, by definition, does not include systematic error or bias.

Accuracy is often expressed inversely in terms of the standard deviation or variance and includes any systematic error or bias. Accuracy includes both the random error of precision and any systematic error. The effect of systematic error on the standard deviation is to inflate it. In the measurement of coal quality for commercial purposes, accuracy expressed in this manner is generally of less interest than is systematic error itself. When systematic error is reduced to a magnitude that is not of practical importance, accuracy and precision can become meaningful parameters for defining truly representative sampling and for interpretation of the results of various test methods.

Estimation of the limits of accuracy (deviation from a true or theoretical value) is not ordinarily attempted in coal analysis. Precision, on the other hand, is determined by means of cooperative test programs. Both *repeatability*, the precision with which a test can be repeated in the same laboratory, usually but not always by the same analyst using the same equipment and following the prescribed method(s), and *reproducibility*, the precision expected of results from different laboratories, are determined. Values quoted in test methods are the differences between two results that should be exceeded in only 5 out of 100 pairs of results, equal to $2\sqrt{2}$ times the standard deviation of a large population of results.

The specification of repeatability and reproducibility intervals, without specification of a statistical confidence level, weakens the precision and accuracy specifications to the extent that this leaves open to question the magnitude of the underlying variance. If, for example, the repeatability interval is never to be exceeded, the variance would have to be zero. From a practical standpoint, this is difficult, if not impossible. Furthermore, the variances (standard deviations) are of direct importance with regard to the details of performing sampling and testing operations because the overall variances can be partitioned into components associated with identifiable sources of variation. This permits assessment of the relative importance of specific details with regard to precision and accuracy. With regard to reproducibility, for example, there is a component of random variance that affects the degree of agreement between laboratories but does not affect the degree of agreement within laboratories (repeatability). Virtually nothing is known about this component of variance except that it exists, and the standard methods do not address this factor directly. However, recognition of it is evidenced in the standard methods by specification of reproducibility intervals that are universally larger than would be accounted for by the variances associated exclusively with the repeatability intervals specified.

In the overall accuracy of results, the sampling variance is but one component, but it is the largest single component. This is a matter of major importance that is frequently missed by the uninitiated. There are test methods (ASTM D-2234; ISO 1988) that describe not only the procedure for the collection of a gross sample of coal but also the method for estimating the overall variance for increments of one fixed weight of a given coal. The precision is such that if gross samples are taken repeatedly from a lot or consignment and one ash determination is made on the analysis sample from each gross sample, 95 out of 100 of these determinations will fall within ±10% of the average of all the determinations. However, under some conditions, this precision may not be obtained, and in terms of performance, the statement should be held in the correct perspective.

At present, when multiseam blended coal samples ranging from 10% by weight mineral matter to as much as 30% by weight mineral matter occur, such precision could result in a corresponding difference as large as 4 to 5% with corresponding differences in the amount of ash that remains after combustion. The response to such concerns is the design of a sampling program that will take into consideration the potential for differences in the analytical data. Such a program should involve acquiring samples from several planned and designated points within (in this case) the coal pile so that allowance is made for changes in the character of the coal as well as for the segregation of the mineral matter during and up to that point in coal's history. That is, the sampling characteristics of the coal play an extremely important role in the application of text methods to produce data for sales.

For coal that is sampled in accordance with standard methods (ASTM D-2234; ASTM D-4596; ASTM D-4916; ASTM D-6315; ASTM D-6518; ISO 13909) and with the standard preparation of the samples for analysis (ASTM D-346; ASTM D-2013), the overall variance of the final analytical data is minimized and falls within the limits of anticipated experimental difference.

1.3 BIAS

The purpose of introducing the term *bias* into coal analysis is to endure the correctness of the analytical data. Understanding the terms *precision* and *bias* as used in quantitative test methods (ASTM D-3670; ASTM D-6300; ASTM D-6708; ASTM E-177) is a necessary part of ensuring the accuracy of the data produced by analytical test methods.

The issue of testing for bias in a coal sampling system (ASTM D-6518; ISO 13 909) is an essential part of coal analysis and is of significant importance (Gould and Visman, 1981). Accordingly, the term *bias* represents the occurrence of a systematic error (or errors) that is (are) of practical importance.

The measurement of systematic error is carried out by taking the differences of replicate results. From a statistical standpoint, to detect a systematic error, it is necessary to reduce the precision limits of the mean to a value less than some multiple of the standard deviation of the differences. To be classified as bias, systematic error must be of a magnitude that is of practical importance. Without proper experimental design, the systematic error may be of a magnitude that is of practical importance because of the various errors. These errors (errors of omission) render the data confusing or misleading and indicate the unreliability of the test method(s).

However, rather than attempt to remove all bias, the aim is to reduce the bias to acceptable levels that do not, in each case, exceed a designated magnitude. Then the test for bias can be designed to confirm the presence of bias when the probability of a bias of that magnitude exists. Indeed, the nature of the problem is such that the absence of bias cannot be proven.

The issues of *relative bias* or *absolute bias* also need consideration. Relative bias is likely to involve comparisons of gross sample results, whereas absolute bias is based on comparison with bias-free reference values and usually involves increment-by-increment comparisons.

The test for bias includes the following essential steps:

1. Pretest inspection
2. Choice of test method specifications
3. Establishment of detailed procedures for conduct of the test method
4. Preliminary test method increment (sample) collection, processing, and analysis
5. Determination of the number of observations required
6. Final increment (sample) collection, processing, and analysis
7. Statistical analysis and interpretation of data

Each variable coal constituent or property to be examined requires assignment of a test method for that variable. As a practical matter, each constituent or property is determined by a test method that can often be viewed on a stand-alone basis. Furthermore, exclusive of moisture, all constituents should be evaluated on a dry basis using a standard size of the coal. Most constituents of coal are affected by errors in size distribution that are associated with size selectivity. Screen tests to obtain size distribution information, particularly in the tails of the size distribution (ISO, 1953), can sometimes prove helpful, but size is not always suitable as a test variable.

Once the data are available, certification of sampling systems as unbiased, without qualification, is insufficient, and certification should also be accompanied by a statement of (1) the mean levels of each variable constituent that prevailed during conduct of the test, (2) the nominal sizing of the coal, and (3) some indication of the preparation (washing) to which the coal has been subjected, since these influence the sampling constants and may affect the magnitude of bias observed.

1.4 REPORTING COAL ANALYSIS

Analyses may be reported on different bases (ASTM D-3180; ISO 1170) with regard to moisture and ash content. Indeed, results that are *as-determined* refer to the moisture condition of the sample during analyses in the laboratory. A frequent practice is to air-dry the sample, thereby bringing the moisture content to approximate equilibrium with the laboratory atmosphere in order to minimize gain or loss during sampling operations (ASTM D-2013; ISO 1988). Loss of weight during air drying is determined to enable calculation on an *as-received basis* (the moisture condition when the sample arrived in the laboratory). This is, of course, equivalent to the *as-sampled basis* if no gain or loss of moisture occurs during transportation to the laboratory from the sampling site. Attempts to retain the moisture at the as-sampled level include shipping in sealed containers with sealed plastic liners or in sealed plastic bags.

Analyses reported on a *dry basis* are calculated on the basis that there is no moisture associated with the sample. The moisture value (ASTM D-3173; ISO 331; ISO 589; ISO 1015; ISO 1018; ISO 11722) is used for converting as determined data to the dry basis. Analytical data that are reported on a *dry, ash-free basis* are calculated on the assumption that there is no moisture or mineral matter associated with the sample. The values obtained for moisture determination (ASTM D-3173; ISO 589) and ash determination (ASTM D-3174) are used for the conversion. Finally, data calculated on an *equilibrium moisture basis* are calculated to the moisture level determined (ASTM D-1412) as the equilibrium (capacity) moisture.

Hydrogen and oxygen reported on the moist basis may or may not contain the hydrogen and oxygen of the associated moisture, and the analytical report should stipulate which is the case because of the variation in conversion factors (Table 1.3). These factors apply to calorific values as well as to proximate analysis (Table 1.4) and to ultimate analysis (Table 1.5).

TABLE 1.3 Conversion Factors of Components Other Than Hydrogen and Oxygen[a]

Given	As-Determined (ad)	As-Received (ar)	Dry (d)	Dry Ash-Free (daf)
As-determined (ad)	—	$\frac{100 - M_{ar}}{100 - M_{ad}}$	$\frac{100}{100 - M_{ar}}$	$\frac{100}{100 - M_{ad} - A_{ad}}$
As-received (ar)	$\frac{100 - M_{ad}}{100 - M_{ar}}$	—	$\frac{100}{100 - M_{ar}}$	$\frac{100}{100 - M_{ar} - A_{ar}}$
Dry (d)	$\frac{100 - M_{ad}}{100}$	$\frac{100 - M_{ar}}{100}$	—	$\frac{100}{100 - A_{d}}$
Dry, ash-free (daf)	$\frac{100 - M_{ad} - A_{ad}}{100}$	$\frac{100 - M_{ar} - A_{ar}}{100}$	$\frac{100 - A_{d}}{100}$	—

Source: ASTM D-3180.

[a] M, percent moisture by weight; A, percent ash by weight. For example, given ad, to find ar, use the formula

$$\text{ar} = \text{ad} \times \frac{100 - M_{ar}}{100 - M_{ad}}$$

TABLE 1.4 Data Derived from Proximate Analysis

	Moisture	Ash	Volatile	Fixed Carbon
Air-dried	8.23	4.46	40.05	47.26
Dry	—	4.86	43.64	51.50
As-received[a]	23.24	3.73	33.50	39.53

[a] Air-dry loss in accordance with ASTM D-2013 = 16.36%.

TABLE 1.5 Data Derived from Ultimate Analysis

Basis	Component (% w/w)							Total
	Carbon	Hydrogen	Nitrogen	Sulfur	Ash	Oxygen[a]	Moisture	(%)
As-determined[b,c]	60.08	5.44	0.88	0.73	7.86	25.01	9.00	100.0
Dry	66.02	4.87	0.97	0.80	8.64	18.70	0.00	100.0
As-received[d]	46.86	6.70	0.69	0.57	6.13	39.05	(29.02)	100.0
As-received[e]	46.86	3.46	0.69	0.57	6.13	13.27	29.02	100.0

[a] By difference.
[b] After air-dry loss (22.00%) in accordance with ASTM D-2013.
[c] Hydrogen and oxygen include hydrogen and oxygen in sample moisture, M_{ad}.
[d] Hydrogen and oxygen include hydrogen and oxygen in sample moisture, M_{ar}.
[e] Hydrogen and oxygen do not include hydrogen and oxygen in sample moisture, M_{ar}.

When hydrogen and oxygen percentages do contain hydrogen and oxygen of the moisture, values on the dry basis may be calculated according to the formulas

$$H_d = (H^1 - 0.1111M^1) \times \frac{100}{100 - M^1} \quad (1.1)$$

$$O_d = (O^1 - 0.8881M^1) \times \frac{100}{100 - M^1} \quad (1.2)$$

where H_d and O_d are the weight percent of hydrogen and oxygen on the dry basis, and H^1 and O^1 are the given or determined weight percents of hydrogen and oxygen, respectively, for the given or determined weight percent of moisture M^1. Rearrangement of these equations to solve for H^1 and O^1 yields equations for calculating moisture containing hydrogen and oxygen contents H^1 and O^1 at any desired moisture level M^1.

The mineral matter (Ode, 1963) in coal loses weight during thermal conversion to ash because of the loss of water from clays, the loss of carbon dioxide from carbonate minerals such as calcite, and the oxidation of pyrite (FeS_2) to ferric oxide (Fe_2O_3). In addition, any chlorine in the coal is converted to hydrogen chloride, but the change in weight may not be significant.

Analyses and calorific values are determined on a mineral-matter-free basis by the Parr formulas (ASTM D-388), with corrections for pyrite and other mineral matter. The amount of pyrite is taken to be that equivalent to the total sulfur of the coal, which despite the potential error has been found to correlate well in studies of mineral matter. The remaining mineral matter is taken to be 1.08 times the weight of the corresponding (iron-oxide-free) ash:

$$mm = 1.08A + 0.55S \quad (1.3)$$

where mm, A, and S are the weight percent of mineral matter, ash, and total sulfur, respectively.

Such data are necessary for calculation of parameters in the classification of coal by rank: dry, mineral-matter-free volatile matter (or fixed carbon) as well as moist, mineral-matter-free gross calorific value. For volatile matter and fixed-carbon data, it is also necessary to assume that 50% by weight of the sulfur is volatilized in the volatile matter test and therefore should not be included as part of the organic volatile matter (nor should the loss from clays and carbonate minerals):

$$FC_{dmmf} = \frac{100(FC - 0.15S)}{100 - (M + 1.08A + 0.55S)} \quad (1.4)$$

$$VM_{dmmf} = 100 - FC \quad (1.5)$$

where FC_{dmmf} and VM_{dmmf} are fixed carbon and volatile matter, respectively, on a dry, mineral-matter-free basis; and FC, M, A, and S are the determined fixed carbon, moisture, ash, and total sulfur, respectively.

In the Parr formula for moist, mineral-matter-free calorific value, the moisture basis used is that of the inherent moisture of the coal in the seam (natural bed moisture, capacity moisture):

$$\text{moist, mineral-matter-free Btu} = \frac{100(\text{Btu} - 50\text{S})}{100 - (1.08\text{A} + 0.55\text{S})} \tag{1.6}$$

where Btu is the calorific value (Btu/lb), A is the ash (% w/w), and S is sulfur (% w/w); all are on the moist (natural bed) basis.

Coal analyses are generally reported in tabular form (Tables 1.4 and 1.5) and the data can be represented graphically as in these EIA coal data from the U.S. Department of Energy:

1. Proximate analysis (see also Table 1.6):

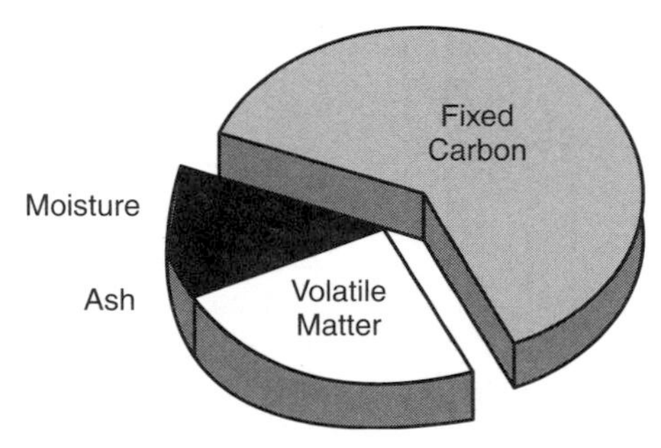

2. Ultimate analysis:

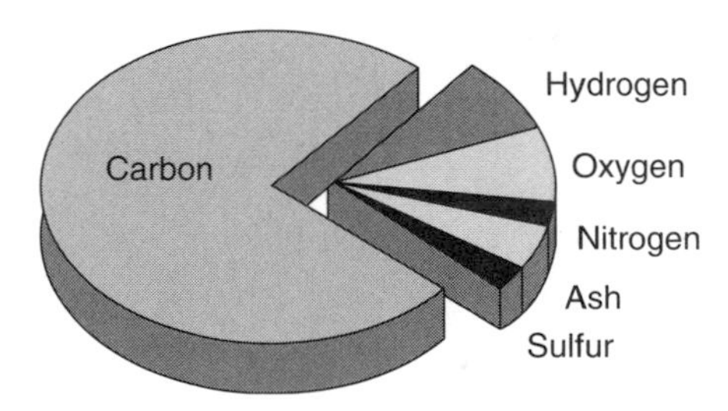

1.5 INTERRELATIONSHIPS OF THE DATA

Just as a relationship exists between the various properties of petroleum with parameters such as depth of burial of the reservoir (Speight, 1999), similar relationships exist for the properties of coal (e.g., Solomon, 1981; Speight, 1994). Variations in hydrogen content with carbon content or oxygen content with carbon content and with each other have also been noted. However, it should be noted that many of the published reports cite the variation of analytical data or test results not with rank in the true sense of the word but with elemental carbon content that can only be approximately equated to rank.

TABLE 1.6 Analytical Specifications for Coal from Selected U.S. Mines[a]

Mine/Owner	Btu/lb	Sulfur (%)	Ash (%)	Moisture (%)	Volatile Matter (%)	Fixed Carbon (%)
Colorado						
Bowie No. 2 Mine	12,000	0.50	9.0	9.0	36.5	49.0
Colowyo Mine	10,460	0.38	5.8	16.9	32.8	44.2
McClane Canyon Mine	11,100	0.50	11.8	8.9	34.4	44.9
Sanborn Creek Mine	12,400	0.52	7.0	6.0	34.6	52.4
Seneca Mine	10,650	0.50	10.1	11.4	34.2	44.3
Twentymile Mine	11,350	0.41	9.3	9.8	35.0	45.9
West Elk Mine	12,000	0.50	9.3	8.5	35.2	47.0
Southern Illinois						
Monterey Mine No. 1	10,300	1.00	8.0	19.0	31.6	41.4
Old Ben 11	11,000	3.00	10.0	13.5	35.7	NA
Rend Lake Mine	NA	NA	NA	NA	NA	NA
Southern Powder River Basin (Wyoming)						
Antelope	8,800	0.22	5.3	26.5	31.8	36.2
Belle Ayr Mine	8,549	0.30	4.6	29.8	31.5	34.1
Black Thunder	8,800	0.34	5.0	27.0	32.0	34.5
Caballo Mine	8,500	0.38	5.1	29.9	NA	NA
Caballo Rojo	8,450	0.32	5.2	30.2	30.6	33.8
Coal Creek	8,350	0.33	5.7	30.0	30.3	31.8
Cordero Mine	8,350	0.37	5.7	30.3	32.0	31.7
Jacobs Ranch	8,686	0.43	5.6	27.9	32.6	33.5
North Antelope–Rochelle Complex	8,800	0.22	4.6	27.6	NA	NA
North Rochelle	8,850	0.20	4.6	26.0	32.1	37.3
Southern Wyoming						
Black Butte Mine	9,700	0.50	7.0	19.5	30.0	42.5
Kemmerer Mine	9,800	0.80	5.0	22.0	34.0	39.5
Medicine Bow Mine	10,400	0.60	7.0	14.0	33.0	NA
Seminole II Mine	11,000	0.65	8.0	11.0	35.0	NA
Shoshone Mine	10,840	0.63	6.3	14.2	37.0	43.0
Utah						
Dugout Canyon Mine	12,000	0.45	9.9	6.3	37.3	46.3
Horizon Mine	12,000	0.65	9.6	9.0	40.1	41.3
Skyline Mine	11,700	0.45	8.5	9.5	40.5	41.5
Sufco Mine	11,400	0.40	8.5	9.5	NA	NA
White Oak Mine	11,750	0.60	7.0	10.0	40.5	42.5
Willow Creek Mine	11,950	0.50	9.0	8.5	41.3	41.2

[a]NA, not available.

Other relationships also exist, such as variations of natural bed moisture with depth of burial as well as the variations in volatile matter content of vitrinite macerals obtained from different depths (Speight, 1994). The latter observation (i.e., the decrease in volatile matter with the depth of burial of the seam) is a striking contrast to parallel observations for petroleum, where an increase in the depth of the reservoir is accompanied by an increase in the proportion of lower-molecular-weight (i.e., more volatile) materials. Similarly, the tendency toward a carbon-rich material in the deeper coal seams appears to be in direct contrast to the formation of hydrogen-rich species (such as the constituents of the gasoline fraction) in the deeper petroleum reservoirs. Obviously, the varying maturation processes play an important role in determining the nature of the final product, as does the character of the source material (Speight, 1994).

Finally, it is also possible to illustrate the relationship of the data from proximate analysis and the calorific value to coal rank.

1.6 COAL CLASSIFICATION

Coal classification is the grouping of different coals according to certain qualities or properties, such as coal type, rank, carbon–hydrogen ratio, and volatile matter. Thus, due to the worldwide occurrence of coal deposits, the numerous varieties of coal that are available, and its many uses, many national coal classification systems have been developed. These systems often are based on characteristics of domestic coals without reference to coals of other countries. However, it is unfortunate that the terms used to describe similar or identical coals are not used uniformly in the various systems.

In the United States, coal is classified according to the degree of metamorphism, or progressive alteration, in the series from lignite (low rank) to anthracite (high rank) (ASTM D-388; Parks, 1963). The basis for the classification is according to yield of fixed carbon and calorific value, both calculated on a mineral-matter-free basis. Higher-rank coals are classified according to fixed carbon on a dry, mineral-matter-free basis. Lower-rank coals are classed according to their calorific values on a moist, mineral-matter-free basis. The agglomerating character is also used to differentiate certain classes of coals.

Thus, to classify coal, the calorific value and a proximate analysis (moisture, ash, volatile matter, and fixed carbon by difference) are needed. For lower-rank coals, the equilibrium moisture must also be determined. To calculate these values to a mineral-matter-free basis, the Parr formulas are used (ASTM D-388).

Thus (Table 1.7), coal with a fixed carbon value in excess of 69% w/w or more, as calculated on a dry, mineral-matter-free basis, are classified according to the fixed-carbon value. Coal with a calorific value below 14,000 Btu/lb, as calculated on a moist, mineral-matter-free basis, is classified according to calorific value on a moist, mineral-matter-free basis, provided that the dry, mineral-matter-free fixed carbon is less than 69%. The agglomerating character is considered for coal

TABLE 1.7 Classification of Coal by Rank[a]

Class	Group	Fixed Carbon Limits (%, Dry, Mineral-Matter-Free Basis)		Volatile Matter Limits (%, Dry, Mineral-Matter-Free Basis)		Calorific Value Limits (Btu/lb, Moist, Mineral-Matter-Free Basis)[b]		Agglomerating Character
		Equal to or Greater Than	Less Than	Greater Than	Equal to or Less Than	Equal to or Greater Than	Less Than	
I. Anthracitic	1. Metaanthracite	98	—	—	2	—	—	Nonagglomerating
	2. Anthracite	92	98	2	8	—	—	
	3. Semianthracite[c]	86	92	8	14	—	—	
II. Bituminous	1. Low-volatile bituminous coal	78	86	14	22	—	—	Commonly agglomerating[e]
	2. Medium-volatile bituminous coal	69	78	22	31	—	—	
	3. High-volatile A bituminous coal	—	69	31	—	14,000[d]	—	
	4. High-volatile B bituminous coal	—	—	—	—	13,000	14,000	
	5. High-volatile C bituminous coal	—	—	—	—	11,500	13,000	
						10,500	11,500	Agglomerating
III. Subbituminous	1. Subbituminous A coal	—	—	—	—	10,500	11,500	Nonagglomerating
	2. Subbituminous B coal	—	—	—	—	9,500	10,500	
	3. Subbituminous C coal	—	—	—	—	8,300	9,500	
IV. Lignite	1. Lignite A	—	—	—	—	6,300	8,300	
	2. Lignite B	—	—	—	—	—	6,300	

Source: ASTM (2004).

[a]This classification does not include a few coals, principally nonbanded varieties, that have unusual physical and chemical properties and that come within the limits of the fixed-carbon or calorific value of the high-volatile bituminous and subbituminous ranks. All of these coals either contain less than 48% dry, mineral-matter-free fixed carbon or have more than 15,500 moist, mineral-matter-free British thermal units per pound.

[b]Moist refers to coal containing its natural inherent moisture but not including visible water on the surface of the coal.

[c]If agglomerating, classify in low-volatile group in the bituminous class.

[d]Coals having 69% or more fixed carbon on the dry, mineral-matter-free basis shall be classified according to fixed carbon, regardless of calorific value.

[e]It is recognized that there may be nonagglomerating varieties in these groups of the bituminous class, and there are notable exceptions in the high-volatile C bituminous group.

TABLE 1.8 International Classification of Hard Coal[a]

Groups[b]			Code Numbers[c]										Subgroups[b]		
	Alternative Group Parameters													Alternative Subgroup Parameters	
Group Number	Free-Swelling Index (Crucible Swelling Number)	Roga Index											Sub-group Number	Dilatometer	Gray–King
3	>4	>45					435	535	635				5	>140	$>G_8$
						334	434	534	634				4	>50–140	G_5–G_8
						333	433	533	633	733			3	>0–50	G_1–G_4
						332 (a, b)	432	532	632	732	832		2	≤0	E–G
2	$2\frac{1}{2}$–4	>20–45				323	423	523	623	723	823		3	>0–50	G_1–G_4
						322	422	522	622	722	822		2	≤0	E–G
						321	421	521	621	721	821		1	Contraction only	B–D

1	1–2	>5–20				212	312	412	512	612	712	812		2	≤0	E–G
						211	311	411	511	611	711	811		1	Contraction only	B–D
0	0–$\frac{1}{2}$	0–5		100		200	300	400	500	600	700	800	900	0	Nonsoftening	A
				a	b											
			Class Numbers[d]											As an indication, the following classes have an approximate volatile matter content of: Class 6: 33–41% Class 7: 33–44% Class 8: 35–50% Class 9: 42–50%		
			0	1		2	3	4	5	6	7	8	9			
Class Parameters	Volatile Matter (Dry, Ash-Free)		0–3	>3–10		>10–14	>14–20	>20–28	>28–33	>33	>33	>33	>33			
				>3–6.5	>6.5–10											
	Calorific Parameter[e]		—	—		—	—	—	—	>13,950	>12,960–13,950	>10,980–12,960	>10,260–10,980			

Source: Adapted with permission from H. H. Lowry, ed., *Chemistry of Coal Utilization*, suppl. vol., John Wiley & Sons, Inc., 1963.

[a] (1) Where the ash content of coal is too high to allow classification according to the present systems, it must be reduced by laboratory float-and-sink methods (or any other appropriate means). The specific gravity selected for flotation should allow a maximum yield of coal with 5–10% ash. (2) 332a . . . > 14–16% volatile matter; 332b . . . > 16–20% volatile matter. (3) Classes determined by volatile matter up to 33% volatile matter and by calorific parameter above 33% volatile matter.

[b] Groups and subgroups are determined by coking properties.

[c] The first figure of the code number indicates the class of the coal, determined by volatile matter content up to 33% VM and by calorific parameter above 33% VM. The second figure indicates the group of coal, determined by coking properties. The third figure indicates the subgroup, determined by coking properties.

[d] Classes are determined by volatile matter up to 33% VM and by calorific parameter above 33% VM.

[e] Gross calorific value on a moist, ash-free basis (30°C, 96% relative humidity, Btu/lb).

with 86% w/w or more dry, mineral-matter-free fixed carbon and for coal with a calorific value between 10,500 and 11,500 Btu/lb, as calculated on a moist, mineral-matter-free basis.

The International Classification of Hard Coals by Type System is based on dry, ash-free volatile matter; calorific value expressed on a moist, ash-free basis; and coking and caking properties. A coal is given a three-figure code number from a combination of these properties (Table 1.8). Coal is first divided into classes 1 to 5, containing coals with volatile matter (dry, ash-free basis) up to 33%. Coal with volatile matter greater than 33% w/w falls into classes 6 to 9, which are separated according to the gross calorific value on a moist, ash-free basis. Although the moist calorific value is the primary parameter for classes 6 to 9, the volatile matter does continue to increase with the rising class number.

The classes of coal are subdivided into groups according to their coking properties, as reflected in the behavior of coals when heated rapidly. A broad correlation exists between the crucible swelling number and the Roga index (ISO methods), and either of these may be used to determine the group number of a coal.

Coals classified by class and by group are further subdivided into subgroups, defined by reference to coking properties. The coking properties are determined by either the Gray–King coke type of assay or the Audibert–Arnu dilatometer test (ISO methods). These tests express the behavior of a coal when heated slowly, as in carbonization.

In the three-figure code number that describes the properties of a coal, the first digit represents the class number, the second the group number, and the third the subgroup number. The international classification accommodates a wide range of coals through use of the nine classes and various groups and subgroups.

Lignite (brown coal) has been classified arbitrarily as coal having a moist, ash-free calorific value below 10,260 Btu/lb. A code number that is a combination of a class number and a group number classifies these coals. The class number represents the total moisture of the coal as mined, and the group number represents the percentage tar yield from dry, ash-free coal (Table 1.9).

TABLE 1.9 International Classification of Hard Coal Using Calorific Value

Group Number	Group Parameter Tar Yield, Dry, Ash-Free (%)	Class Parameter; Total Moisture, Ash-Free[a] (%)					
		Class 10 0–20%	Class 11 20–30%	Class 12 30–40%	Class 13 40–50%	Class 14 50–60%	Class 15 60–70%
40	>25	1040	1140	1240	1340	1440	1540
30	20–25	1030	1130	1230	1330	1430	1530
20	15–20	1020	1120	1220	1320	1420	1520
10	10–15	1010	1110	1210	1310	1410	1510
00	10 and less	1000	1100	1200	1300	1400	1500

Source: Reprinted with permission from H. H. Lowry, ed., *Chemistry of Coal Utilization*, suppl. vol., John Wiley & Sons, Inc., 1963.

[a] The total moisture refers to freshly mined coals. Moist, ash-free basis: 30°C and 96% relative humidity.

1.7 THE FUTURE

Coal analysis has, by convention, involved the use of wet analysis or the use of typical laboratory bench-scale apparatus. This trend continues and may continue for another decade or two. But the introduction of microprocessors and microcomputers in recent years has led to the development of a new generation of instruments for coal analysis as well as the necessary calibration of such instruments (ASTM D-5373). In particular, automated instrumentation has been introduced that can determine moisture, ash, volatile matter, carbon, hydrogen, nitrogen, sulfur, oxygen, and ash fusion temperatures in a fraction of the time required to complete most standard laboratory bench procedures.

Several such instruments have been developed for the simultaneous determination of carbon, hydrogen, and nitrogen in various samples. Of course, basic requirements for the instruments are that they provide for the complete conversion of carbon, hydrogen, and nitrogen in coal to carbon dioxide, water vapor, and elemental nitrogen, and for the quantitative determination of these gases in an appropriate gas stream.

A disadvantage of some of the instrumental methods for determining carbon, hydrogen, and nitrogen is the small sample size used in the analysis. On the best of days, a typical sample size for some of the instruments might be 1 to 3 mg, but the accuracy of the system might be questioned. Other systems that use 100-mg samples may be preferred, provided that effluents do not flood or overpower the system and overcome the detection equipment. However, the larger sample size does increase the probability that the sample is representative of the quantity of coal being analyzed.

Most methods used by the new analytical all-in-one instruments are empirical, and the accuracy of the results is highly dependent on the quality and suitability of the standards used to standardize the instruments.

REFERENCES

ASTM. 2004. *Annual Book of ASTM Standards*, Vol. 05.06. American Society for Testing and Materials, West Conshohocken, PA. Specifically:

ASTM D-121. Standard Terminology of Coal and Coke.

ASTM D-346. Standard Practice for Collection and Preparation of Coke Samples for Laboratory Analysis.

ASTM D-388. Standard Classification of Coals by Rank.

ASTM D-1412. Standard Test Method for Equilibrium Moisture of Coal at 96 to 97 Percent Relative Humidity and 30°C.

ASTM D-2013. Standard Practice of Preparing Coal Samples for Analysis.

ASTM D-2234. Standard Practice for Collection of a Gross Sample of Coal.

ASTM D-3173. Standard Test Method for Moisture in the Analysis Sample of Coal and Coke.

ASTM D-3174. Standard Test Method for Ash in the Analysis Sample of Coal and Coke from Coal.

ASTM D-3180. Standard Practice for Calculating Coal and Coke Analyses from As-Determined to Different Bases.

ASTM D-3670. Standard Guide for Determination of Precision and Bias of Methods of Committee D22.

ASTM D-4596. Standard Practice for Collection of Channel Samples of Coal in a Mine.

ASTM D-4916. Standard Practice for Mechanical Auger Sampling.

ASTM D-5373. Standard Test Methods for Instrumental Determination of Carbon, Hydrogen, and Nitrogen in Laboratory Samples of Coal and Coke.

ASTM D-6300. Standard Practice for Determination of Precision and Bias Data for Use in Test Methods for Petroleum Products and Lubricants.

ASTM D-6315 (withdrawn 2003). Standard Practice for Manual Sampling of Coal from Tops of Barges.

ASTM D-6518. Standard Practice for Bias Testing a Mechanical Coal Sampling System.

ASTM D-6708. Standard Practice for Statistical Assessment and Improvement of the Expected Agreement Between Two Test Methods that Purport to Measure the Same Property of a Material.

ASTM E-177. 2004. Standard Practice for Use of the Terms Precision and Bias in ASTM Test Methods.

BS. 2003. *Methods for Analysis and Testing of Coal and Coke*. BS 1016. British Standards Institution, London.

Gluskoter, H. J. 1975. In *Trace Elements in Fuel*, S. P. Babu (Editor). Advances in Chemistry Series 141. American Chemical Society, Washington, DC, pp. 1–22.

Gluskoter, H. J., Shimp, N. F., and Ruch, R. R. 1981. In *Chemistry of Coal Utilization*, 2nd Suppl. Vol., M. A. Elliott (Editor). Wiley, Hoboken, NJ, p. 411.

Gould, G., and Visman, J. 1981. In *Coal Handbook*, R. A. Meyers (Editor). Marcel Dekker, New York, p. 19.

ISO. 1953. *Hard Coal—Size Analysis by Sieving*.

ISO. 2003. *Standard Test Methods for Coal Analysis*. International Organization for Standardization, Geneva, Switzerland. Specifically:

ISO 331. Determination of Moisture in the Analysis of Coal.

ISO 589. Determination of the Total Moisture of Hard Coal.

ISO 1015. Determination of the Moisture Content of Brown Coals and Lignites.

ISO 1018. Determination of the Moisture Holding Capacity of Hard Coal.

ISO 1170. Calculation of Analyses to Different Bases.

ISO 1988. Sampling of Hard Coal.

ISO 11722. Hard Coal: Determination of Moisture in the General Analysis Test Sample by Drying in Nitrogen.

ISO 13909. Mechanical Sampling: Parts 1, 2, 3, 4, 7, and 8.

Karr, C. K., Jr. (Editor). 1978. *Analytical Methods for Coal and Coal Products*, Vols. 1 and 2. Academic Press, San Diego, CA.

Lowry, H. H. (Editor). 1963. *Chemistry of Coal Utilization*, Suppl. Vol., Wiley, Hoboken, NJ.

Montgomery, W. J. 1978. In *Analytical Methods for Coal and Coal Products*, Vol. 1, C. K. Karr, Jr. (Editor). Academic Press, San Diego, CA, Chap. 6.

Ode, W. H., 1963. In *Chemistry of Coal Utilization*, Suppl. Vol., H. H. Lowry (Editor). Wiley, Hoboken, NJ, Chap. 5.

Parks, B. C. 1963. In *Chemistry of Coal Utilization*, Suppl. Vol., H. H. Lowry (Editor). Wiley, Hoboken, NJ, Chap. 5, pp. 29–34.

Patrick, J. W., and Wilkinson, H. C. 1978. In *Analytical Methods for Coal and Coal Products*, Vol. 2, C. K. Karr, Jr. (Editor). Academic Press, San Diego, CA, Chap. 29.

Rees, O. W. 1966. *Chemistry, Uses, and Limitations of Coal Analysis*. Report of Investigations 220. Illinois State Geological Survey, Urbana, IL.

Smith, K. L., and Smoot, L. D. 1990. *Prog. Energy Combust. Sci.*, 16:1.

Solomon, P. R. 1981. In *New Approaches in Coal Chemistry*, B. D. Blaustein, B. C. Bockrath, and S. Friedman (Editors). Symposium Series 169. American Chemical Society, Washington, DC, p. 61.

Speight, J. G. 1994. *The Chemistry and Technology of Coal*, 2nd ed. Marcel Dekker, New York.

Speight, J. G. 1999. *The Chemistry and Technology of Petroleum*, 3rd ed. Marcel Dekker, New York.

van Krevelen, D. W. 1961. *Coal*. Elsevier, Amsterdam.

Vorres, K. S. 1993. *Users' Handbook for the Argonne Premium Coal Sample Program*. Argonne National Laboratory. Argonne, IL; National Technical Information Service, U.S. Department of Commerce, Springfield, VA.

2 Sampling and Sample Preparation

For homogeneous materials, sampling protocols are relatively simple and straightforward, although caution is always advised lest overconfidence cause errors in the method of sampling as well as introduce extraneous material. On the other hand, the heterogeneous nature of coal (Speight, 1994, and references cited therein) complicates the sampling procedures. In fact, apart from variations in rank (Chapter 1), coal is often visibly heterogeneous and there is strong emphasis on the need to obtain representative samples for testing and analysis (Gould and Visman, 1981).

Thus, the variable composition of coal offers many challenges to analysts who need to ensure that a sample under investigation is representative of the coal. Indeed, the substantial variation in coal quality and composition from the top to the bottom of the seam, from side to side, and from one end to the other, within an unmined bed offers challenges that are perhaps unprecedented in other fields of analytical chemistry: hence the issues that arise during drilling programs designed to determine the size and extent of a coal bed or coal seam. This variability in coal composition and hence in coal quality is often significantly, and inadvertently, increased by mining, preparation, and handling.

Transportation (by belt, rail, or truck) can initiate (due to movement of the coal) processes that result in size and density segregation. Thus, variations from one side of a conveyor belt to the other, from side-to-side, end-to-end, and top-to-bottom locations in individual cars or trucks, and between one location and another in a coal pile, must be anticipated (ASTM D-346; ASTM D-2234; ASTM D-4182; ASTM D-4702; ASTM D-4915; ASTM D-4916; ASTM D-6315; ASTM D-6518; ASTM D-6543; ISO 1988). Therefore, the challenge in sampling coal from a source or shipment is to collect a relatively small portion of the coal that accurately represents the composition of the coal. This requires that sample increments be collected such that no piece, regardless of position (or size) relative to the sampling position and implement, is collected or rejected selectively. Thus, the coal sample must be representative of the composition of the whole coal (i.e., coal in a pile or coal in a railcar or truck) as represented by the properties or quality of the sample.

Handbook of Coal Analysis, by James G. Speight
ISBN 0-471-52273-2

Optimization of coal sampling is a function of the many variable constituents of coal. The effect of fineness on the combustion of pulverized coal is dramatic, and the special problems associated with collection of an unbiased sample of pulverized coal need to be addressed (ASTM D-197). *Operating samples* are often collected from the coal streams to power plants on a regular basis not only for determination of heat balance but also to document compliance with air pollution emission regulations.

Thus, to test any particular coal, there are two criteria that must be followed for a coal sample (1) ensure that the sample is a true representative of the bulk material, and (2) ensure that the sample does not undergo any chemical or physical changes after completion of the sampling procedure and during storage prior to analysis. In short, the reliability of a sampling method is the degree of perfection with which the identical composition and properties of the entire body of coal are obtained in a sample. The reliability of the storage procedure is the degree to which the coal sample remains unchanged, thereby guaranteeing the accuracy and usefulness of the analytical data.

2.1 SAMPLING

Samples submitted for chemical and physical analyses are collected for a variety of reasons, but the collection of each sample should always conform to certain guidelines. The application of precise techniques in sample collection helps to ensure that data from each analysis performed on the samples will be useful. For interpretations and comparisons of elemental compositions of coal beds to be valid, the samples must be collected so that they are comparably representative of the coal bed. Such interpretations and comparisons should never be based on data from different types of samples (Swanson and Huffman, 1976; Golightly and Simon, 1989).

Thus, sampling plays a role in all aspects of coal technology. The usual example given is the determination of coal performance in a power plant. However, an equally important objective relates to exploration and sampling of coal reserves as they exist in the ground. The issues in this case relate not only to determining the extent of the coal resource but also to the quality of the coal so that the amount may be determined. Thus, sampling in connection with exploration is subject to (1) the location, (2) the spacing of the drilled holes, (3) the depth from which the sample is taken, and (4) the size of core drills used. These criteria must be taken into consideration when assessing the quality and quantity of coal in the deposit being explored.

More to the current point, reliable sampling of a complex mixture such as coal is difficult, and handling and quite often the variations in coal-handling facilities make it difficult to generate fixed rules or guidelines that apply to *every* sampling situation. Proper collection of the sample involves an understanding and consideration of the minimum number and weight of increments, the particle size distribution of the coal, the physical character and variability of the constituents of coal, and the desired precision of the method. Thus, preliminary to any laboratory testing of coal, it is imperative that a representative sample of the coal be obtained in as reproducible and repeatable a manner as possible. If not, data derived from the most carefully conducted analysis are meaningless.

A *gross sample* of coal is a sample that represents a quantity, or *lot,* of coal and is composed of a number of increments on which neither reduction nor division has been performed (ASTM D-2234). The recommended maximum quantity of coal to be represented by one gross sample is 10,000 tons [usually, the tonnage shipped in a *unit train*: 100 cars, each of which contains 100 tons of coal (although a unit train may now contain 110 cars or more)]. Mineral matter content (often incorrectly designated as ash content) is the property most often used in evaluating sampling procedures. The density segregation of the mineral matter speaks to the movement of the coal particles relative to each other during transportation. Environmentally, sulfur content has also been applied in the evaluation of sampling procedures.

The sampling procedures (ASTM D-346; ASTM D-2234; ASTM D-4702; ASTM D-4915; ASTM D-4916; ASTM D-6315; ASTM D-6518; ASTM D-6543) are designed to give a precision such that if gross samples are taken repeatedly from a lot or consignment and prepared according to standard test methods (ASTM D-197; ASTM D-2013) and one ash determination is made on the analysis sample from each gross sample, the majority (usually specified as 95 out of 100) of these determinations will fall within ±10% of the average of all the determinations. When other precision limits are required or when other constituents are used to specify precision, defined special-purpose sampling procedures may need to be employed.

Thus, when a property of coal (which exists as a large volume of material) is to be measured, there usually will be differences between the analytical data derived from application of the test methods to a *gross lot* or *gross consignment* and the data from the *sample lot*. This difference (the *sampling error*) has a frequency distribution with a mean value and a variance. *Variance* is a statistical term defined as the mean square of errors; the square root of the variance is more generally known as the *standard deviation* or the *standard error of sampling*.

Recognition of the issues involved in obtaining representative samples of coal and minimization of the *sampling error* has resulted in the designation of methods that dictate the correct manner for coal sampling (ASTM D-346; ASTM D-2234; ASTM D-4702; ASTM D-4915; ASTM D-4916; ASTM D-6315; ASTM D-6518; ASTM D-6543; ISO 1988; ISO 2309).

Every sampling operation consists of either extracting one sample from a given quantity of material or of extracting from different parts of the lot a series of small portions or *increments* that are combined into one gross sample without prior analysis; the latter method is known as *sampling by increments*. In fact, the number of riffling stages required to prepare the final sample depends on the size of the original *gross lot*. Nevertheless, it is possible by use of these methods to reduce an extremely large consignment (which may be on the order of tons, i.e., several thousand pounds) to a representative sample (1 pound or less) that can be employed as the sample for the application of laboratory test methods.

The precision of sampling is a function of the size of increments collected and the number of increments included in a gross sample, improving as both are increased, subject only to the constraint that increment size not be small enough to cause selective rejection of the largest particles present. Recognition of this was evidenced in the specification of minimum number and weight of increments in coal sampling (ASTM D-2234). The manner in which sampling is performed as it relates to the precision of the sample thus depends on the number of increments collected from all parts of the lot and the size of the increments. In fact, the number and size of the increments are operating variables that can, within certain limits, be regulated by the sampler.

Considerations pertinent to the procurement of a representative sample of coal from a gross lot include the following:

1. The *lot* of coal must first be defined (e.g., a single truck, about 20 tons; a barge, about 1500 tons; a unit train, about 10,000 tons; or a ship cargo, about 100,000 tons).
2. The number of increments (e.g., the number of shovelfuls required to constitute the gross sample, which is usually 200 to 500 lb) must be established.
3. For raw, dirty, or poorly cleaned coal, the minimum number of increments is 35.
4. For thoroughly cleaned coal (i.e., maximum practical reduction of ash and sulfur), the minimum number of increments is 15.
5. The precision (ASTM D-2234) is based on one analytical determination falling within one-tenth of the true value 95 times out of 100. To reduce this error by one-half, four times as many gross samples must be used.
6. The weight per increment varies according to the top size of the coal.
7. Increments must be spaced systematically. Stationary sampling employs a grid system, which may be a simple left front–middle center–right rear grid for samples from a railroad car or a surveyed grid system to take samples from a storage pile.
8. Additionally, increments taken from a coal storage pile take into account any variations in the depth of the pile.
9. Increments from a moving coal stream are often collected on a preset interval of time by a mechanical sampling device. The opening of the device must be sufficient to accommodate a full stream cut in both directions without disturbing the coal.

Stream sampling and *flow sampling* are terms usually reserved for the collection of sample increments from a free-falling stream of coal as opposed to the collection of increments from a motionless (stopped) conveyor belt. Coal that passes from one belt to another at an angle tends to become segregated because of the momentum caused by density and particle size differences, with a predominance of coarse particles on one side and a predominance of fine particles on the other side.

Sampling at rest consists of acquiring a coal sample when there is no motion. In such instances, it may be difficult, if not impossible, to ensure that the sample is truly representative of the gross consignment. An example of coal being sampled at rest is when samples are taken from railcars (*car-top sampling*), and caution is advised both in terms of the actual procedure and in the interpretation of data. Again, some degree of segregation can occur as the coal is loaded into hopper cars. In addition, heavy rainfall can cause the moisture content of the coal to be much higher at the top and sides of a railcar than at the bottom. Similarly, the onset of freezing conditions can also cause segregation of the moisture content.

Sampling error is the difference that occurs when the property of the representative sample is compared to the true, unknown value of the gross lot or consignment. The sampling error has a frequency distribution with a mean value and a variance. *Variance* is a statistical term defined as the mean square of errors. Its square root is the more broadly known statistic called the *standard deviation*, or *standard error*, of sampling. Sampling error can thus be expressed as a function of the sampling variance or sampling standard deviation, each of which, in turn, is directly related to the material and the specifics of sample collection.

One aspect of coal sampling materials that has been employed when it is suspected that the gross coal sample (the coal pile or the coal in a railcar after transportation) is nonrandomly distributed is known as *stratified sampling* or *representative sampling*. The procedure consists of collecting a separate sample from each stratum of the gross material lot and determining the properties from each sample so obtained. Incremental sampling has been considered to be a form of stratified sampling in which the strata are imaginary because there is no physical boundary between the imaginary strata, and any such segregation is identified with the portions from which the individual increments are collected. The *within-strata* and *between-strata variances* are a function of the size and number of increments.

Preparation plant performance testing and routine quality control in mining operations and preparation plants require sampling coal both in situ and at various stages of processing following removal from the bed. Other than *channel sampling* for sampling coal in situ, and the sampling of coal slurries, the sampling techniques for quality control purposes and preparation plant are necessary. However, assessing preparation plant performance may require complex sampling programs for the sampling of many coal streams with widely different sampling properties involving the collection of sample increments for which the timing has to be tightly coordinated. Such sampling almost always depends on manual sampling with a variety of sampling implements, often in locations with difficult if not inadequate access.

Storage of laboratory coal samples for subsequent analysis is also a part of proper sample handling. Long-term storage without change is achieved by placing the samples in a plastic bag containing dry ice, sealing them tightly in glass beakers, and storing them under vacuum. Normally, oxidation and deterioration of 60-mesh laboratory samples stored in air increase with decreasing particle size and decreasing rank of coal.

In summary, the precision of sampling improves with the size of each of the increments collected and with the number of increments included in a gross sample; and manual sampling involves the principle of ideal sampling insofar as every particle in the entire mass to be sampled has an equal opportunity to be included in the sample.

2.1.1 Manual Sampling

There are two considerations involved with the principle of manual sampling (that every particle in the entire mass to be sampled have an equal opportunity to be included in the sample): (1) the dimensions of the sampling device, and (2) proper use of the sampling device. The opening of the sampling device must be two to three times the top size of the coal to meet sampling method (ASTM D-2234) requirements, and design criteria have been established for several types of hand tools that can be used for manual sampling (Figure 2.1). The main considerations are that the width is not less than the specified width and the device must be able to hold the minimum specified increment weight without overflowing.

One particular method of sampling (ASTM D-6883) that relates to the standard practice for manual sampling of stationary coal from railroad cars,

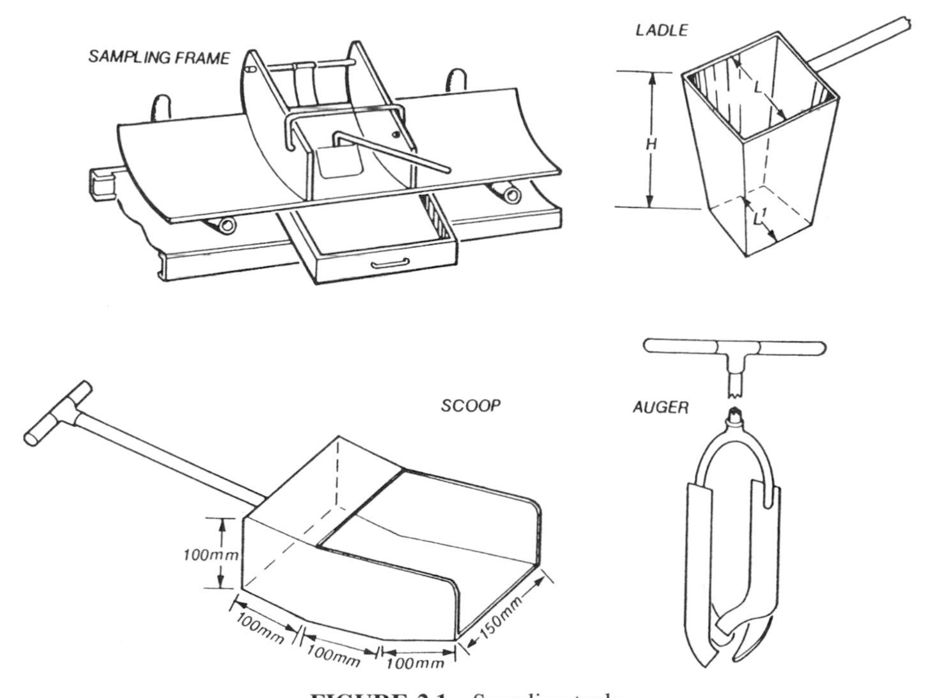

FIGURE 2.1 Sampling tools.

barges, trucks, or stockpiles (ASTM D-6315; ASTM D-6610). These procedures described in this method are to be used to provide gross samples for estimating the quality of the coal. The gross samples are to be crushed, divided, and further prepared for analysis (ASTM D-2013).

The practices described by the method provide instructions for sampling coal from beneath the exposed surface of the coal at a depth (approximately 24 in., 61 cm) where drying and oxidation have not occurred. The purpose is to avoid collecting increments that are significantly different from the majority of the lot of coal being sampled due to environmental effects. However, samples of this type do not satisfy the minimum requirements for probability sampling and, as such, cannot be used to draw statistical inferences such as precision, standard error, or bias. Furthermore, this method is intended for use only when sampling by more reliable methods that provide a probability sample is not possible.

Systematic spacing of increments collected from a stopped belt is accepted universally as the reference method of sampling that is intrinsically bias-free. *Stationary sampling*, that is, sampling coal at rest in piles, or in transit in trucks, railcars, barges, and ships, suffers decreased reliability to an indeterminate degree.

Sampling from coal storage piles (*sampling at rest*) is not as simple as may be perceived and can have serious disadvantages. For example, coal in conical-shaped piles suffers segregation effects that result in fines predominating in the central core (ASTM D-5192) as well as a gradation of sizes down the sides of the pile from generally fine material at the top of the pile to coarser coal at the base of the pile. If at all possible, coal piles should be moved before sampling, which, in turn, determines how the coal is sampled.

Where it is not possible to move a pile, there is no choice but to sample it *as is*, and the sampling regime usually involves incremental spacing of samples over the entire surface. The reliability of the data is still in doubt. However, without any attempt at incremental spacing of the sample locations, any sample taken directly from an unmoved storage pile is a *grab sample* that suffers from the errors that are inherent in the structure of the pile as well as in the method by which the sample is obtained.

Alternatively, sample acquisition from large coal piles can be achieved by core drilling or by use of an auger, or the coal can be exposed at various depths and locations (by means of heavy equipment such as a bulldozer) so that manual sampling can be performed. A wide variety of devices are available for machine sampling (mechanical sampling) and include flow-through cutters, bucket cutters, reciprocating hoppers, augers, slotted belts, fixed-position pipes, and rotating spoons (Figures 2.2 to 2.4). A major advantage of these systems is that they sample coal from a moving stream (ASTM D-6609).

There are numerous situations where coal must be sampled at rest despite the potential for compromising the reliability of the sample(s) acquired. A major problem with sampling coal at rest is that an inevitable and unknown degree of segregation will prevail, and it is not possible to penetrate all parts of the mass such that every particle has an equal opportunity to be included in the sample.

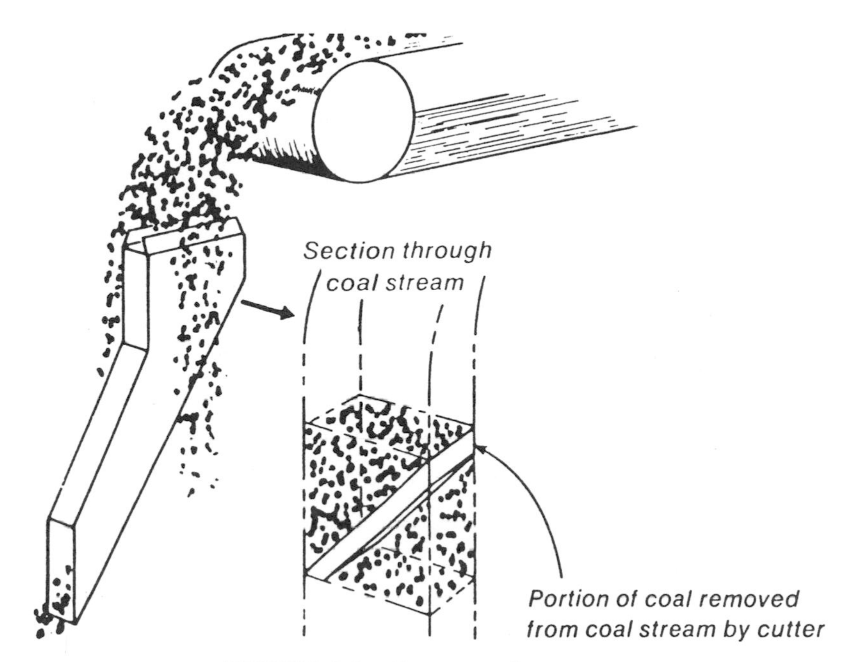

FIGURE 2.2 Cross-cut primary cutter.

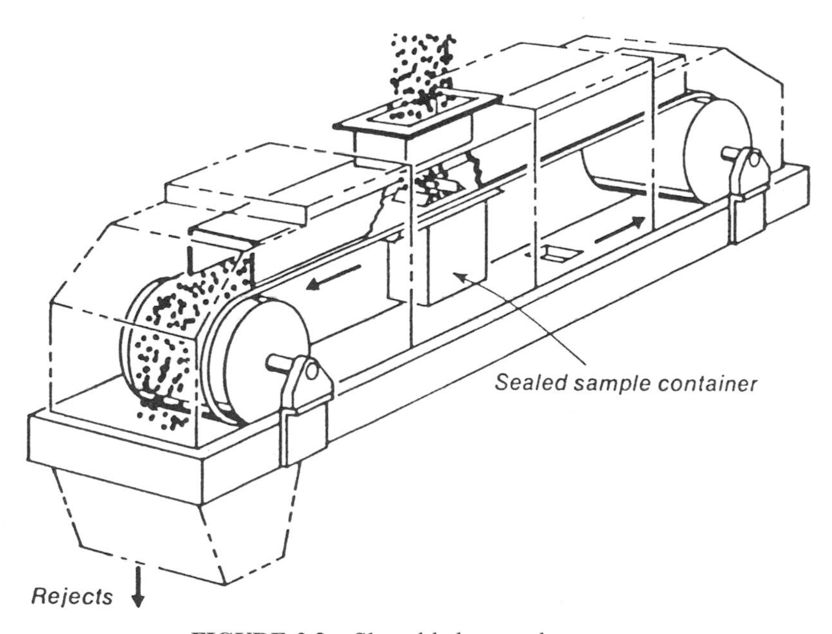

FIGURE 2.3 Slotted-belt secondary cutter.

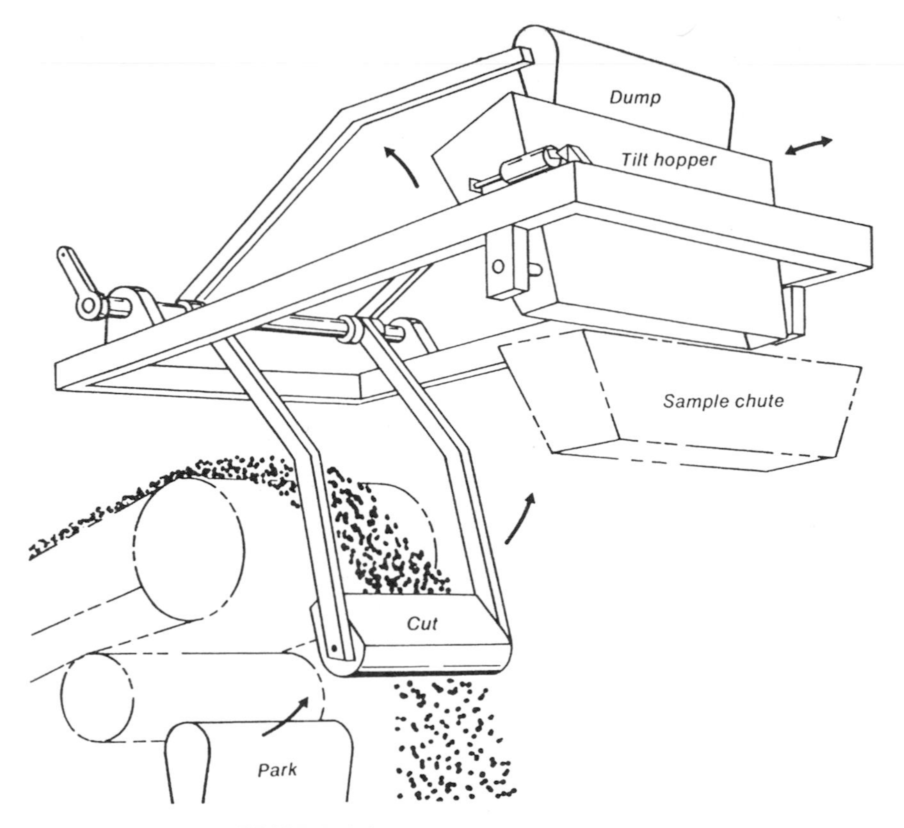

FIGURE 2.4 Swing-arm primary cutter.

The commonest situation where coal must be sampled at rest arises where the coal has to be sampled from railcars. The alternatives for sampling from hopper cars are, top, bottom, and a combination of the two. However, as coal is loaded into hopper cars, it suffers segregation related to the loading process that is not necessarily obvious, and the degree to which it will affect sample results is unpredictable. Sometimes the segregation is clearly evident, such as when cars are loaded from a stream that enters the car from the side, causing the large pieces to be shot to the far side while fines remain at or close to the near side. Having made such a statement, segregation can also occur in a more discreet or subtle manner, and any differences in texture and appearance are not always clearly visible.

In addition, when a significant amount of surface moisture is present, some will begin migrating downward immediately, resulting in a substantially higher moisture content at the bottom of the car than at the top. Furthermore, the difference in moisture at different levels may become more pronounced as time passes, owing to the effects of evaporation and precipitation. Heavy rainfall can cause the moisture content to be much higher at the top and sides than at the bottom, and if freezing conditions prevail (as they often do during Wyoming winters), the outcome will be even more obvious.

In *car-top sampling*, only the coal near the top surface has the potential to be included in the sample, thereby violating the basic tenet of obtaining a representative sample. Thus, the uncertainties regarding the accuracy of the results are increased and any conclusions drawn from the data are highly suspect. Therefore, if car-top sampling is a necessity, the increments must not be collected predominantly from any given location relative to the dimensions of the railcar. Furthermore, if the railcars vary substantially in size, the number of increments per car should be varied proportionately.

An alternative operation to sample the coal is to employ *bottom sampling*, in which coal is sampled as it is discharged from the bottom of hopper cars. Since the coal is *sampled in motion*, bottom sampling is considered to be an improvement over car-top sampling.

Stream sampling (*flow sampling*) is the sampling of coal in motion, usually from one part of the plant to another. However, increment collection must involve cutting across the full stream. The collection of increments from the sides of a moving belt is sometimes loosely called *stream* or *flow sampling*, and this terminology therefore should not be accepted as assurance that increments were collected from a free-falling stream. Such procedures are less reliable because the increments collected are subject to analytical bias caused by any segregation of the coal (including the mineral matter) that has occurred ion the conveyer belt. In fact, coal larger than 1 in. top size, or coal heaped to a depth of more than about 8 in. on a conveyor belt, exhibits a tendency to segregate. Coal that passes from one belt to another at an angle invariably becomes more segregated, with a greater predominance of coarse particles on the far side and a greater predominance of fine particles on the near side.

When the falling stream is more than 1 ft thick or is more than about 2 ft wide, the forces involved tend to be greater than can be resisted with handheld equipment, and simple mechanical devices are often useful. A pivoted scoop with the necessary mechanical advantage is useful, and slide gates (a dropout section at the bottom of a scraper conveyor) or a flop gate in a vertical chute are other possibilities. In the event that such alternatives are not feasible, partial stream cuts are permissible, but the reliability of the sampling is reduced. To combat any reduction in the reliability, partial stream cuts need to be made systematically at different points in the stream so that all parts of the coal stream are represented proportionately.

Despite the potential advantages of the procedure, sampling coal in motion may suffer from disadvantages such as (1) it is not always possible to penetrate the full depth of the coal cascading out of the car; (2) attempts to penetrate the stream result in sample scoop overflow; (3) increment collection is limited to the exposed surface at the sides of the car; (4) moisture is often higher at the sides of the car than for the entire contents of the car; (5) flow rates are highly variable; and (6) disproportionate amounts of coarse coal are often collected because the coarse particles segregate and roll down the exposed surface.

2.1.2 Mechanical Sampling

A wide variety of mechanical devices are now in use and include flow-through cutters, bucket cutters, reciprocating hoppers, augers, slotted belts, fixed-position pipes, and rotating spoons. These systems typically collect the primary increments and perform at least part of the sample preparation by crushing and dividing it down to the 4- or 8-mesh stage of reduction specified (ASTM D-2013).

Conventional design of most mechanical sampling systems for large tonnages of coal use some form of *cross-stream primary cutter* to divert the primary increments from the main stream of coal (Figure 2.2). A major advantage of these systems is that they sample coal from a moving stream, and most of them satisfy the principle that every particle in the entire mass has an equal opportunity to be included in the primary increments.

One of the more innovative primary cutter designs that reduce the component of impact velocity perpendicular to the direction of coal flow in the plane in which the cutter is moving is the *swing-arm cutter* (Figure 2.4). This design achieves this goal by passing through the stream at an obtuse angle instead of at a right angle and can be moved at higher speeds than a cross-stream cutter without causing analytical bias that is associated with disturbance to the coal flow.

A *pipe sampler* is an off-the-belt sampler that collects increments from within the stream cross section by means of one or two pipes. The sampling pipes are mounted at an obtuse angle on a horizontal axle positioned at right angles to the direction in which the coal flows and increments are collected through an orifice that is located in the bottom wall of the sampling pipe.

A *rotary car dumper system* (*tube sampler*) consists of several large-diameter tubes, each with one or two openings that are attached to a rotary car dumper. The openings are located in the path of the coal as it is discharged from the railcar. If, as has been claimed, the tubes collect coal predominantly from the top and far side of the car, the method is susceptible to bias because of the potential segregation of the coal constituents.

A *spoon sampler* consists of one or more pipes, arranged like the spokes of a wheel. Openings located at the tips collect the sample as the device is rotated through coal on a moving belt. This machine can be designed to collect very small primary increments, but the spoon pipes may overflow during increment collection and the sample may be of questionable reliability.

An *auger drill* is also used as a sampling device for penetrating a stationary mass of coal and withdrawing material from its interior. At least two specialized auger-sampling machines, designed for sampling from trucks, are commercially available. One of these uses a 10-in. auger and is intended for sampling uncrushed run-of-mine coal, and a truck-mounted portable version is also used for sampling from railcars.

2.2 SAMPLE PREPARATION

Once a gross sample has been taken, it is reduced in both particle size and quantity to yield a *laboratory sample*. This aspect is known as *coal preparation*.

Sample preparation (ASTM D-2013) includes drying (in air), as well as crushing, dividing, and mixing a gross sample to obtain a sample that is ready for analysis. As written, this test method covers the reduction and division of gross or divided samples up to and including the individual portions for laboratory analysis. Reduction and division procedures are prescribed for coals of the following groups: group A, which includes coals that have been cleaned in all sizes and allows lower-weight laboratory samples to be retained than in group B, which includes all other coals, including unknown coals.

Two processes of sample division and reduction are covered: (1) procedure A, in which manual riffles are used for division of the sample and mechanical crushing equipment for reduction of the sample, and (2) procedure B, in which mechanical sample dividers are used for division of the sample and mechanical crushing equipment for reduction of the sample. A third process that is, in reality, a combination of procedures A and B may be used at any stage.

Other standards are used to collect the gross sample, and one test method allows for one division of the gross sample before crushing. The mass and top size of the gross or divided sample collected using these guides and practices are usually too large for chemical or physical testing. However, any bias in the gross or divided sample before adherence to this practice will remain in the final sample resulting from use of this method. Therefore, the standard to be used to collect the gross sample should be selected carefully. Often, the sample is collected, reduced, and divided (one or more times) by use of a mechanical sampling system, and the remaining sample may be divided further on-site to facilitate transport to the laboratory, where further reduction and division probably occurs before analysis. But division and reduction of a sample may occur at more than one location. Samples are reduced and divided to provide an analysis sample, but some test methods require a sample of different mass or top size. This method can be adapted to provide a sample of any mass and *size consist* (particle size distribution) from a gross or divided sample up to, and including, an analysis sample.

Since moisture losses are a perpetual problem, part of the procedure may include weighing, air drying, and reweighing (ASTM D-3302) before crushing and dividing. This provides a sample in which the moisture loss during the preparation procedure has been determined and has provided a stabilized sample that is not subject to further moisture loss. Thus, sample preparation is not just simply a matter of dividing a gross sample into manageable or usable increments. The task must also be accomplished in a manner that produces an unbiased sample ready for analysis.

When executed improperly, sample preparation is the source of the second-largest component of the overall variance of sampling and analysis. Although it is not specified as a requirement, it is generally recognized that the variance of sample division and analysis is not more than 20% of the total variance of sampling, division, and analysis (ASTM D-2013). The particle size distribution (size consist) of the laboratory sample depends on its intended use in the laboratory and the nature of the test methods to be applied. The minimum allowable weight of the sample at any stage of reduction depends on the size consist and the degree of precision desired (ASTM D-2013).

Many issues, including (1) loss or gain of moisture, (2) improper mixing of constituents, (3) improper crushing and grinding, and (4) oxidation of coal, may arise during the sampling and sample preparation processes. To minimize moisture contamination, all standard methods include an air-drying stage in the preparation of the analysis sample. In this manner, all subsequent handling and analysis are made on a laboratory sample of relatively stable moisture content. Be that as it may, all collecting, handling, reducing, and division of the gross sample should be performed as rapidly as possible and in as few steps as possible to guard against further moisture loss or gain in the ambient laboratory conditions.

The distribution of mineral matter in coal presents problems for crushing, grinding, and uniform mixing at each step of the sampling procedure. The densities of the various coal constituents cause segregation, especially if there is a wide particle size distribution. Thus, crushing and/or grinding coal from a large particle to a very small particle should involve a reasonable number of steps that are based on the starting particle size and nature of the coal. At the same time, too many handling steps increase the exposure of the coal to air and increase the chance of moisture variation and coal oxidation. On the other hand, attempting to crush, grind, or pulverize coal from a large to a small particle size in one operation tends to produce a wide range of particle sizes and a high concentration of very fine particles.

Coal is susceptible to oxidation at room temperature, and like the potential for change in the moisture content, the adverse effects of oxidation must be considered in sampling, coal preparation, and coal storage. Again and where possible, the coal preparation steps should be done rapidly, and in as few steps as possible, to minimize oxidation of the coal. Sample containers used should have airtight lids to guard against moisture loss and exposure of the coal to air. In addition, the containers should be selected to hold only the desired amount of sample and to leave a minimum of airspace. Even then, analysis of a sample should be carried out as soon as possible after it is received. Prolonged storage before analysis is often disadvantageous.

The effect of fines content on the combustion of pulverized coal is quite dramatic (Field et al., 1967; Essenhigh, 1981), and the problems associated with collection of an unbiased sample of pulverized coal need attention (ASTM D-197). Operating samples are often collected from the feedstocks to power plant boilers on a shift or daily basis for calculation of heat balances and operating efficiencies. Another objective of operating samples is to document compliance with air pollution emission regulations based on fuel composition.

2.3 WASHABILITY

Coal washing is a process by which mineral matter is removed from coal by the use of any one of several washing processes to leave the coal as near mineral-free as is required by the buyer or by legislation. Mineral matter occurs in coal as in two clearly defined forms, intrinsic mineral matter and extrinsic mineral matter. *Intrinsic mineral matter* is present in intimate association with the pure coal substance itself and originates from inorganic material essential to the growth of the vegetable matter from which the coal was formed originally. Owing to its physical condition, such mineral matter cannot be separated from the coal substance by physical means, but since it seldom exceeds 1% by weight of the coal substance, it does not lead to undue difficulties with ash when the coal is burned in the normal way.

Extrinsic mineral matter, which is purely adventitious, is derived from the roof and floor of the coal seam and from any noncoal or inorganic material that may be associated with the seam itself. It consists generally of pieces of stone, clay, and shale together with infiltrated inorganic salts that have become deposited in the natural fissures in the coal seam (e.g., pyrite, ankeritic material). Such material can be reduced very much in amount by suitable methods of coal cleaning and, indeed, may be separated from the coal completely, provided that it can be broken apart from coal particles.

In simplest terms, the gravimetric separation of light and heavy fractions is used to accomplish coal washing. This involves one or more float–sink tests carried out on predesignated size fractions into which the coal is divided by screening. Solutions or suspensions of different specific gravities are used for the separations. This is performed repetitively with higher and higher specific gravity solutions, separating the size fraction into gravity fractions. The coal is generally of lower density than the mineral matter, and these gravity fractions thus generally exhibit higher and higher ash content. Each size fraction and each gravity fraction is dried, weighed, and analyzed. The analysis is usually made for ash and sulfur content but is frequently extended to include heating value and other variables.

In the test method for determining the washability characteristics of coal (ASTM D-4371), the need for a standard procedure to conduct washability analyses that will serve as an aid to technical communication to coal suppliers and purchaser is recognized. This test method standardizes procedures utilized for performing washability analyses, the data from which can be used for interpreting preparation plant efficiency, for determining preparation plant design, and for determining the potential recovery and quality of coal reserves. This test method describes procedures for determining the washability characteristics of coarse- and fine-coal fractions. Each sample being tested can have more than one coarse-coal size fraction and more than one fine-coal size fraction (ASTM D-4749 provides a test method for the sieve analysis of coal).

In this test method (ASTM D-4371), the specific gravity fractions are obtained by subjecting the material being studied to a series of solutions, each with a discrete specific gravity, that cover the range of specific gravities in question. These solutions are obtained by the mixing of various organic liquids that are relatively inert toward the coal. The distribution, as determined by the analysis, is affected by the physical condition of the sample subjected to the washability analysis (e.g., the moisture content and the size content of the material). Furthermore, this method may not be the most technically correct test method to determine washability characteristics of low-rank coals because of problems relative to the loss of moisture through drying during sample preparation and analysis. Methods that are directly applicable to low-rank coal are not yet available.

The testing procedure consists of washing the coal in a hand jig or subjecting it to a series of various organic liquids of increasing specific gravity. Although certain closely sized portions of a coal sample may show washing characteristics quite different from those of other portions of the same sample, it is customary in practice to have to examine the characteristics of a sample containing all size ranges simply as a whole. A *hand jig*, consisting of a Henry tube, is suitable for moderately small graded coals (Figure 2.5). The jig, or moving portion of the washer, is a brass tube about 30 in. long and 4-to-6 in. in diameter, fitted with handles at the top and having a gauze bottom (about 40 mesh) held in position by two setscrews and a reinforcing disk, all of which form a loose bottom to the jig. The entire apparatus is immersed in an outer vessel containing water. When jigging is complete, the samples are set aside to air-dry in a dust-free atmosphere and then weighed. Each sample is ground, the percentage of

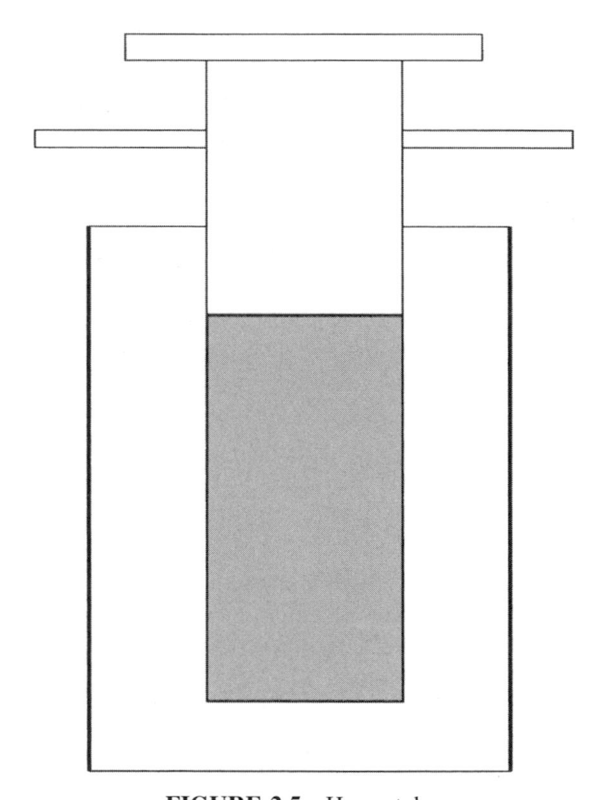

FIGURE 2.5 Henry tube.

moisture and ash is determined, and from the results a set of washability curves can be constructed.

Determination of the washability characteristics of coal by the float and sink (float–sink) method can be applied to coal of any particle size provided suitably large vessels to hold the larger lumps are available. Air-dried coal, not dry coal, should be used since the separation depends partly on the difference in specific gravity of the clean coal and dirt particles, and the specific gravity, in turn, is dependent on the moisture content of the coal. If the coal is dried before the test is carried out, the conditions will then differ from those in commercial washers, and the results will be at variance with those obtained in practice.

The solutions used are generally mixtures of organic liquids such as benzene (specific gravity 0.88), toluene (0.87), carbon tetrachloride (1.60), and bromoform (2.90), and for most purposes a range of liquids from specific gravity 1.20 to 1.60 by increments of 0.05 is adequate. If the coal contains an appreciable quantity of large pieces, it is advantageous to separate the sample into two portions on a 1-in. screen and wash the two fractions separately. The larger fraction may suitably be washed in increments in a tall glass cylinder, the floats removed as soon as separation is complete, and the sinks removed from the bottom of the cylinder from time to time as the occasion demands.

It is customary to start with the lightest liquid and then to treat the sinks with the liquid with the next-higher specific gravity. The various fractions are then set aside to air-dry and are mixed with the corresponding fractions obtained by washing the coal under the 1- or 2-in. screen size. Care should be exercised when taking the original sample to avoid such a wide size range as to preclude obtaining a thoroughly representative sample.

If the coal to be washed has no corresponding fraction over 1 in., the sample is air-dried and a small representative portion is removed for determination of moisture and ash. Weighed portions of the coal are then treated with liquid of specific gravity 1.20, and after the initial separation of floats and sinks has taken place, allowed to stand until separation is complete. The floats are filtered from the solution through a filter paper in a Buchner funnel and set aside to air-dry. The sinks are recovered in a similar manner and allowed to dry. The test is repeated with the other portions of the sample. All the floats are collected together, as are the sinks. The air-dried sinks are subjected to a liquid of specific gravity 1.25, the floats are separated as before, the sinks are subjected to the next liquid, and so on, until finally, the sinks at, say, 1.60 are obtained. All samples are air-dried and weighed (combined with their corresponding samples of coal over 1 in. if necessary), and all or a portion of each sample is crushed for determination of moisture and ash.

If bromoform is used, it is best to wash the fractions after filtration with a little methyl alcohol to dissolve the bromoform since the high boiling point of bromoform [151°C (304°F)] makes it difficult to remove completely by air drying. Furthermore, although it has been recommended that solutions of inorganic salts (e.g., calcium chloride, zinc chloride) in water may be used to replace the more expensive organic solvents, it is almost impossible to remove these salts from various fractions, even by much washing with water, with the result that the ash figures become less reliable.

With decreasing particle size, the practical feasibility of gravity separation diminishes because the efficiency of separation is reduced. Particularly in low-volatile coals, a substantial proportion of the coal may be lost in these fines if other means are not applied for separating them from the slime (clay) and recovering them. *Froth flotation* is one technique that is used for this purpose.

In principle, *froth flotation* consists of bubbling air through a suspension of fine coal and water to which various chemical agents have been added to improve the processes. The separation occurs by reason of a preferential physical attachment of air bubbles to the coal. The coal particles float to the top and are removed. The procedure is sometimes repeated, reprocessing the float fraction several times to simulate multistage full-scale froth flotation.

Selective flotation of one mineral in preference to another is accomplished by depositing on the grains of the first a film of a suitable reagent that will promote flotation of that mineral. Reagents that condition the surface of a material in this manner are known as *collectors*, those inducing the formation of a stable froth in which the material floated can collect are termed *frothers*, and reagents having the property of inhibiting the flotation of one or more constituents of a mixture are called *depressants*.

Coal cleaning by froth flotation is applied essentially to those coals that are inherently soft and are generally obtained in a size range so small as to be difficult to treat by methods of coal cleaning that are based primarily on the differences in specific gravity between the clean coal particles and those of the rejected material. Specially cleaned coking coals for use in the manufacture of coke for electrode carbons are usually prepared by froth flotation, the collectors employed being generally creosote oil fractions from coal tar, essential oils, pine oil, and so on.

The mechanical processes involved in *jig washing* and *float–sink tests* are those that yield a series of increments of the original coal with increasing mineral matter content. As a consequence, it is often found that the integrated ash for the entire sample differs appreciably from the value obtained by direct determination. This difference is the result of the ash being prepared under two sets of conditions. First, each separate increment is ashed (i.e., combusted so that the only residue is mineral ash) under conditions that differ from increment to increment because the mineral matter present on coal can change from one increment to the other. Second, the whole of the coal is ashed to give a certain reproducible ash yield obtainable from any average sample of the original coal. The divergence will be greatest for those coals containing a large percentage of pyrite sulfur, especially if this is also associated with a high alkali or alkaline earth content. The divergence can be minimized, but not eliminated entirely, if the ashing processes are carried out in two stages, as recommended in the proximate analysis.

The net result of the washability test methods is to subdivide coal into fractions having progressively decreasing mineral matter ash content and progressively increasing specific gravities. From the data furnished by the tests, the washability characteristics of the sample can be ascertained. These characteristics are presented most conveniently by construction of *washability curves* for the coal under examination.

Normally, three curves are constructed for each coal: (1) the *instantaneous ash curve* or *coal characteristic curve*, which gives the ash content of any of the individual layers into which the coal has, or can be, separated; (2) the *integrated ash curve for clean coal*, which gives the percentage yield of cleaned coal having an ash content of a certain amount and the ash content of a mixture of any number of consecutive layers of cleaned coal from the top down to a fixed layer, together with the corresponding yield expressed as a percentage of the original coal; and (3) the *integrated ash curve for dirt*, which gives the percentage yield of reject material having an ash content greater than a certain amount or the ash content of a mixture of any number of consecutive layers of dirty coal from the bottom layer up to a fixed layer, together with the yield expressed as a percentage of the original coal.

If the coal contains fusain, this maceral will appear in the lightest fraction. If much is present, its relatively high mineral matter content may have such a weighted effect on the ash yield of the first fraction as to make it greater than the ash yielded by the second, and indeed subsequent, fractions. As a result, the characteristic curve may develop a pronounced curvature up and to the right at the top extremity. Fusain and its disturbing effect should not, however, be ignored because if the sample of coal washed is representative of the main bulk of coal to be washed, the effect of the presence of the fusain must be acknowledged.

REFERENCES

ASTM, 2004. *Annual Book of ASTM Standards*, Vol. 05.06. American Society for Testing and Materials, West Conshohocken, PA. Specifically:

ASTM D-197. Standard Test Method for Sampling and Fineness Test of Pulverized Coal.

ASTM D-346. Standard Practice for Collection and Preparation of Coke Samples for Laboratory Analysis.

ASTM D-2013. Standard Practice of Preparing Coal Samples for Analysis.

ASTM D-2234. Standard Practice for Collection of a Gross Sample of Coal.

ASTM D-3302. Standard Test Method for Total Moisture in Coal.

ASTM D-4182. Standard Practice for Evaluation of Laboratories Using ASTM Procedures in the Sampling and Analysis of Coal and Coke.

ASTM D-4371. Standard Test Method for Determining the Washability Characteristics of Coal.

ASTM D-4702. Standard Guide for Inspecting Crosscut, Sweep-Arm, and Auger Mechanical Coal-Sampling Systems for Conformance with Current ASTM Standards.

ASTM D-4749. Standard Test Method for Performing the Sieve Analysis of Coal and Designating Coal Size.

ASTM D-4915. Standard Guide for Manual Sampling of Coal from Tops of Railroad Cars.

ASTM D-4916. Standard Practice for Mechanical Auger Sampling.

ASTM D-5192. Standard Practice for Collection of Coal Samples from Core.

ASTM D-6315. Standard Practice for Manual Sampling of Coal from Tops of Barges.

ASTM D-6518. Standard Practice for Bias Testing a Mechanical Coal Sampling System.

ASTM D-6543. Standard Guide to the Evaluation of Measurements Made by On-Line Coal Analyzers.

ASTM D-6609. Standard Guide for Part-Stream Sampling of Coal.

ASTM D-6610. Standard Practice for Manual Sampling Coal from Surfaces of a Stockpile.

ASTM D-6883. Standard Practice for Manual Sampling of Coal from Railroad Cars, Barges, Trucks or Stockpiles.

BS. 2003. *Methods for Analysis and Testing of Coal and Coke*. BS 1016. British Standards Association, London.

Essenhigh, R. H. 1981. In *Chemistry of Coal Utilization*, 2nd Suppl. Vol., M. A. Elliott (Editor). Wiley, Hoboken, NJ, Chap. 19.

Field, M. A., Gill, D. W., Morgan, B. B., and Hawksley, P. G. W. 1967. *Combustion of Pulverized Coal*. British Coal Utilization Research Association, Leatherhead, Surrey, England.

Golightly, D. W., and Simon, F. O. 1989. *Methods for Sampling and Inorganic Analysis of Coal*. U.S. Geological Survey Bulletin No. 1823. U.S. Department of the Interior, Washington, DC.

Gould, G., and Visman, J. 1981. In *Coal Handbook*, R. A. Meyers (Editor). Marcel Dekker, New York, p. 19.

ISO. 2003. *Standard Test Methods for Coal Analysis*. International Organization for Standardization, Geneva, Switzerland. Specifically:

ISO 1988. Sampling of Hard Coal.

ISO 2309. Sampling of Coke.

Speight, J. G. 1994. *The Chemistry and Technology of Coal*, 2nd ed. Marcel Dekker, New York.

Swanson, V. E., and Huffman, C., Jr. 1976. *Guidelines for Sample Collecting and Analytical Methods Used in the U.S. Geological Survey for Determining the Chemical Composition of Coal*. U.S. Geological Survey 735. U.S. Department of the Interior, Washington, DC.

3 Proximate Analysis

The proximate analysis of coal is *an assay* of the *moisture, ash, volatile matter,* and *fixed carbon* as determined by series of prescribed or standard test methods. It was developed as a simple means of determining the distribution of products obtained when the coal sample is heated under specified conditions. By definition, the proximate analysis of coal separates the products into four groups: (1) moisture; (2) volatile matter, consisting of gases and vapors driven off during pyrolysis; (3) fixed carbon, the nonvolatile fraction of coal; and (4) ash, the inorganic residue remaining after combustion. On occasion, proximate analysis may include ash fusion temperature and the free-swelling index of coal or the swelling properties of coal (ASTM D-720; ISO 8264) (Chapter 7).

The proximate analysis of coal is presented as a group of test methods (ASTM D-3172; ASTM D-3173; ASTM D-3174; ASTM D-3175; ASTM D-5142; ISO 1171) that has been used widely as the basis for coal characterization in connection with coal utilization. These analyses are in contrast to the ultimate analysis of coal, which provides information about the elemental composition.

The standard test method for proximate analysis (ASTM D-3172) covers the methods of analysis associated with the proximate analysis of coal and coke and is, in fact, a combination of the determination of each of three of the properties and calculation of a fourth. Moisture, volatile matter, and ash are all determined by subjecting the coal to prescribed temperature levels for prescribed time intervals. The losses of weight are, by stipulation, due to loss of moisture and, at the higher temperature, loss of volatile matter. The residue remaining after ignition at the final temperature is called *ash*. Fixed carbon is the difference of these three values summed and subtracted from 100. In low-volatile materials such as coke and anthracite coal, the fixed-carbon value equates approximately to the elemental carbon content of the sample.

The term *volatile matter content* (of coal) is actually a misnomer, insofar as the majority of the volatile matter is the volatile product of the thermal decomposition of coal through the application of high temperatures. The extent to which the more volatile smaller molecules of coal (Vahrman, 1970) add to this is dependent on the coal and should be determined by nondestructive methods such as extraction by solvent(s). Relative yields and boiling-point profiles provide the extent to which natural molecules contribute to the volatile matter without any influence from high-temperature cracking.

Handbook of Coal Analysis, by James G. Speight
ISBN 0-471-52273-2

The final results of the proximate analysis of coal (ASTM D-3172; ASTM D-3173; ASTM D-3174; ASTM D-3175; ASTM D-5142; ISO 562) are usually reported to the first decimal place; any subsequent figures have little or no significance. The final report of the analysis should always contain the results on a basis of air-dried coal (i.e., coal in its most stable condition and in which it was analyzed), but for purposes of classification or comparison it is often necessary to convert to another basis, such as *dry coal, dry, ash-free coal*, or *as-received* coal.

3.1 MOISTURE CONTENT

There are several sources of the water found in coal. The vegetation from which coal was formed had a high percentage of water that was both physically and chemically bound, and varying amounts of this water were still present at various stages of the coalification process. But the overall result of the continuation of the coalification process was to eliminate much of the water, particularly in the later stages of the process, as is evident from a comparison of the moisture contents of different ranks of coal, from lignite to anthracite (Table 3.1). Water is present in most mines and circulates through most coal seams. After mining, many coals are washed with water during preparation for market and are then subject to rain and snow during transportation and storage. All of these sources contribute to the moisture in coal and to the problems associated with measurement of this moisture.

The total moisture in coal is the determination of the moisture (in all forms except water of crystallization of the mineral matter) that resides within the coal matrix. In fact, moisture (or water) is the most elusive constituent of coal to be measured in the laboratory. The moisture in coal ranges from 2 to 15% by weight in bituminous coal to nearly 45% by weight in lignite.

The standard methods of determining the amount of moisture in coal include a variety of test methods designed to differentiate between the various types of

TABLE 3.1 Composition and Property Ranges for Various Ranks of Coal

	Anthracite	Bituminous	Subbituminous	Lignite
Moisture (%)	3–6	2–15	10–25	25–45
Volatile matter (%)	2–12	15–45	28–45	24–32
Fixed carbon (%)	75–85	50–70	30–57	25–30
Ash (%)	4–15	4–15	3–10	3–15
Sulfur (%)	0.5–2.5	0.5–6	0.3–1.5	0.3–2.5
Hydrogen (%)	1.5–3.5	4.5–6	5.5–6.5	6–7.5
Carbon (%)	75–85	65–80	55–70	35–45
Nitrogen (%)	0.5–1	0.5–2.5	0.8–1.5	0.6–1.0
Oxygen (%)	5.5–9	4.5–10	15–30	38–48
Btu/lb	12,000–13,500	12,000–14,500	7500–10,000	6000–7500
Density (g/mL)	1.35–1.70	1.28–1.35	1.35–1.40	1.40–1.45

moisture in the coal: (1) ASTM D-1412 (ISO 1018) for determination of the equilibrium moisture of coal at 96 to 97% relative humidity and 30°C, (2) ASTM D-2961 for determination of the total moisture of coal reduced to number 8 (2.38-mm) top sieve size (limited-purpose method), (3) ASTM D-3173 for determination of the moisture in the analysis sample of coal and coke, and (4) method D-3302 for determination of the total moisture in coal. In addition to these test methods, the method of preparing coal samples for analysis (ASTM D-2013) gives directions for air-drying coal samples. It has been suggested that the best technique is to determine loss during drying in air (ASTM D-3302; ISO 11722) followed by codistillation of moisture with xylene. Alternatively, moisture can be determined in an oven (at a fixed temperature) that is swept with dry nitrogen or another inert gas (ISO 589).

No absolute method for determining the true moisture content of coal by heating has been devised because of the wide variance in the temperatures at which different forms of moisture (including surface, inherent, chemically combined moisture in the coal) and water of hydration in clay minerals are liberated. There are, however, several test methods of an empirical nature of methods for determining the amount of moisture in coal. In one method the weight loss of coal was measured after heating to 106°C (222°F) in a flow of dry, oxygen-free nitrogen. Although two other methods (weight loss in air and direct measurement of moisture released in a nitrogen atmosphere), also at 106°C (222°F), yielded results within acceptable tolerances, those performed in a nitrogen atmosphere gave consistently higher results because heating in air sometimes oxidizes coal and offsets the moisture loss through addition of oxygen. Measuring indirect weight loss after heating in nitrogen at 130°C (266°F) for 30 minutes with a flow rate of 90 oven-volume changes per hour is, perhaps, a more convenient method.

Moisture loss can (and often does) occur during grinding and/or pulverization and is believed to be related to the types and amounts of banded (maceral) ingredients in the coal. Losses are least for vitrain and increase in order of vapor pressures for moisture (e.g., vitrain < clarain < durain < fusain).

The role of water in coal and the quantitative measurement of water are complicated because water is present within the coal matrix in more than one form (Allardice and Evans, 1978). Thus, the total moisture includes both the surface moisture and the residual moisture remaining in the sample after determining the air-dry loss (ASTM D-3302). Thus,

$$M = \frac{R(100 - ADL)}{100} + ADL \tag{3.1}$$

where M is the total moisture (% by weight), R the residual moisture (% by weight), and ADL the air-drying loss (% by weight).

To determine the total moisture, either an analysis sample can be prepared from the moisture sample, or the regular analysis sample can be used for this purpose, provided that the moisture analysis is performed on the analysis sample within a prescribed time after the air-dry sample is prepared. When separate analysis samples are used for moisture and for the other determinations that are made, moisture determinations must be made on both. The moisture obtained on the regular analysis sample is then used only for the purpose of calculating results to the *dry basis*. The moisture determined on the analysis sample prepared from the separate moisture sample is combined with the air-dry loss to determine the total moisture, and the total moisture is then used for calculation of *as-received basis* results.

Methods for determination of the total moisture in coal have been placed into the following categories: (1) thermal methods that often include distillation methods; (2) a desiccator method; (3) distillation methods, which often include extraction and/or solution methods; (5) chemical methods; and (6) electrical methods.

Most common tests for moisture involve a thermal drying procedure, usually at a temperature a few degrees above the boiling point of water; the moisture released on heating is measured either directly or indirectly. Another thermal method that is often used involves moisture determination by measuring the weight loss of a coal sample on heating in various atmospheres. If the coal is susceptible to oxidation, as most low-rank coals with high moisture content are, the heating should be done in an inert atmosphere.

The desiccator method involves weight loss (of the coal) and/or the gain in weight of a desiccant in a weighing tube that adsorbs moisture from the atmosphere above the coal. This test method is probably the more accurate method, since the tube absorbs only water, although other gases, such as methane, may be evolved. Moisture may also be determined as the loss in weight when coal is heated to various temperatures (with the atmosphere and pressure variable) or by determination of the weight gain of a vessel containing a desiccant through which passes the volatile materials evolved when the coal is heated.

The distillation method of moisture determination requires collection and determination of the water evolved from the coal when the sample is heated in a boiling solvent that is itself immiscible with water. The solution and extraction methods require either solvent extraction of the water from the coal (followed by subsequent determination of the water content of the solvent) or use of a standard reagent that will exhibit differences in concentration by virtue of the water in the coal. A nonthermal solvent method of determining moisture involves the use of an extraction procedure in which the coal is shaken with a solvent that extracts the water from the coal. The degree of change in some physical property of the solvent, such as density, is then used as a measure of the water extracted.

The chemical methods of water determination invoke the concepts of direct titration of the water or a chemical reaction between the water and specific reagents that causes the evolution of gases; the water is determined by measurement of the volume produced. Chemical methods used for determining moisture also include (1) application of the Karl Fischer method of determining water content, and (2) reaction of quicklime with water in coal and subsequent measurement of the heat generated by the reaction.

Electrical methods of measuring coal moisture involve determination of the capacitances or resistances or the dielectric constant of coal from which the water content can be determined. These methods have been used by industry, particularly for moving streams of coal.

Each method has advantages or disadvantages that must be considered prior to the application or acceptance of the method. Furthermore, applicability of any one method is dependent not only on accuracy (as well as the reproducibility of that accuracy) but also on the applicability of that method to the entire range of coal types.

Generally, the moisture content of coal is determined by measuring the weight loss of a sample (expressed as a percentage of the sample) that has been comminuted to pass through a 250-μm (60-mesh) sieve (ASTM D-3173; ISO 331; ISO 687). The sample (about 1 g) is maintained under controlled conditions ($107 \pm 3°C$, $225 \pm 5°F$) in an inert atmosphere for 1 hour.

After thermal drying methods, distillation methods are the next most commonly used. In these procedures, coal is heated in a liquid that has a boiling point higher than that of water and is immiscible with it. Xylene, toluene, or a petroleum fraction of a selected boiling range are the liquids normally used. The distilled vapors are condensed in a graduated tube, and the volume of water is measured after the two liquids separate (BS, 2003; ISO 348; ISO 1015). This type of method is considered particularly advantageous for use with low-rank coal, since air is excluded from the coal and any error due to oxidation is minimized. This is also a direct method of measuring moisture; consequently, there is no error due to the loss of other gases.

Thus, the procedure is particularly applicable for use with high-moisture, low-rank coal, which may also be sensitive to aerial oxygen. For example, with lignite (and other low-rank coals), oxidation during the 1-hour heating period can, conceivably, cause a weight increase that albeit small, is significant enough to reduce the amount of moisture found in the coal. In addition, although the moisture as determined will, indeed, be mostly moisture, it may include some adsorbed gases, whereas some strongly adsorbed moisture will not be included. Caution is also necessary to ensure that the coal sample is not liable to thermal decomposition at the temperature of the moisture determination.

In all methods for total moisture determination, the total moisture may be determined using a single-stage or two-stage method in which the as-received sample is air-dried at approximately room temperature and the residual moisture is determined in the sample (ASTM D-3302; ISO 589). The air-drying step reduces the water in the sample to an equilibrium condition, depending on the laboratory humidity and, as a result, minimizes any potential changes in the moisture content that might occur when the sample is prepared for further analysis. As a result, the single-stage method does not involve the intermediate stage and is classified as a limited-purpose method, recognizing that certain coals will give results that reflect varying levels of oxidation. High-rank coal may be affected somewhat less than lower-rank coal by this treatment.

Determination of the residual moisture (ASTM D-3173) involves another determination, and the data are used for calculating other analytical results to the dry basis. It is also used in conjunction with the air-dry loss, when it is determined, to calculate results to the as-received basis. On those occasions when *very wet coal* that is in comparatively large pieces of coal is in hand, one unique method is available for determination of the moisture content. In the test method, the sample (in a container) is transferred to a clean, weighed metal tray and reweighed immediately. The coal is then spread evenly over the surface of the tray by means of a spatula (or other instrument) and should be left on the tray until the coal can be dislodged from the spatula blade by tapping and allowed to air-dry in a well-ventilated room free of dust. Reweighing is performed periodically until two successive weightings differ by less than 0.1% by weight of the weight of the original sample. At that time, the final weight of the air-dried coal is determined by weighing the sample, container, and tray together; the cleaned container is weighed and the loss in weight (due to moisture lost by air drying) is expressed as a percentage of the original sample. The entire sample can then be crushed and ground (with reduction in bulk if necessary) to give a sample possessing the necessary size qualities, which, without further drying, can be subjected to a determination of moisture content as specified in the test for proximate analysis.

Another test method that can be used, determination of the moisture content of *moderately dry coal*, can, in fact, be used for all coal samples that can be handled without appreciable loss of moisture. The principle of the method is that the coal is heated with an inert liquid that does not mix with water and which has a boiling point greater than that of water so that, on distillation, the water is carried over with the vapor from the liquid and, on condensation, separates from it in a graduated receiver. The Dean and Stark apparatus (Figure 3.1) or one of its many modifications can be used for this purpose.

In the procedure (ISO 331), coal (whose weight depends on moisture content) is weighed into a flask and dry toluene is added. The apparatus is assembled and, after bringing the toluene to the boil, boiling is continued until no more water separates out in the receiver. When no change in the volume of the water takes place over a specified time interval (usually 15 minutes), any moisture adhering to the inner tube of the condenser is transferred to the receiver by washing down with dry toluene, and the receiver is allowed to cool until it reaches room temperature. Once all of the water has been collected in the bottom of the receiver, read the lowest level of the water–toluene meniscus, apply any calibration correction necessary, and express the quantity of water obtained as a weight percent of the coal. A well-defined cutoff point for moisture release is achieved without significant oxidation of the coal, and for this reason, the use of toluene or xylene distillation methods for brown coal is recommended.

This test method is applicable to coal of any size but is limited by the diameter of the neck of the flask. The apparatus must be scrupulously clean [usually by a mixture of sulfuric acid and potassium dichromate, after which the apparatus is washed with distilled water and dried in an oven at about 110°C (230°F)]. The condenser should be cleaned periodically in the same manner, and the receiver should be capable of collecting up to 10 mL of water in graduations of 0.1 or 0.05 mL, with a calibration for the water–toluene meniscus.

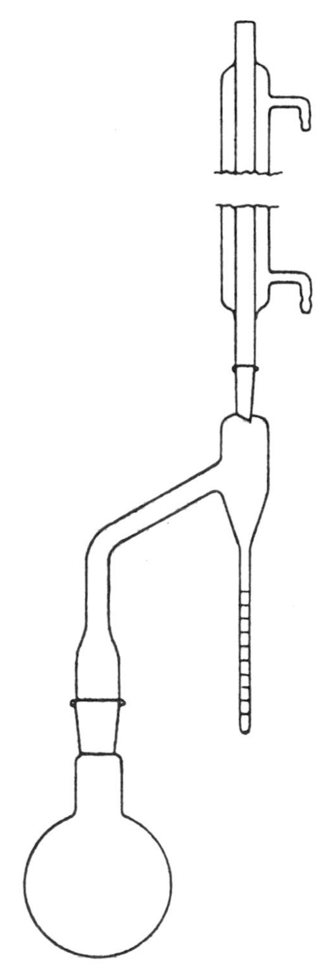

FIGURE 3.1 Dean and Stark apparatus.

Other methods that have been used in determination of the amount of moisture in coke include (1) extraction of coke using anhydrous methanol and the addition of calcium hydride with coal constituents (CaH_2) from which the amount of released heat is measured, and (2) extraction of coke using anhydrous dioxan and measurement of the refractive index of the solution to determine its water content. Application of these methods to coal may be subject to error because of (1) interaction of the coal (in contrast to coke) with the calcium hydride, leading to chemical errors, and (2) the influence of extractable constituents on the refractive index of the dioxan.

It is often difficult to apply indirect weight-loss procedures to the determination of moisture content in low-rank coal that has a high moisture content and is susceptible to oxidation. Some ovens may be inadequate for determining the amount of moisture in brown coal, but it must be ascertained that moisture does not continue to be released from the coal after prolonged periods of heating, leaving no sharp moisture-release cutoff point. On this basis, water is perhaps best removed from low-rank coal by distillation with toluene in which a well-defined moisture cutoff is reached (ISO 1015).

The continuous measurement of moisture in coal has been accomplished by (1) electrical conductivity, (2) dielectric constant, (3) microwave attenuation, (4) neutron scattering, (5) nuclear magnetic resonance, (6) infrared, and (7) thermal conductivity (Hampel, 1974).

3.1.1 Test Method Protocols

Moisture determination (ASTM D-3173; ASTM D-3302) depends on the extent of sample preparation and the condition of the coal sample. The entire procedure for determining the total moisture in coal, after collecting the gross sample, begins with preparing the sample for analysis (ASTM D-2013). If the gross sample is sufficiently dry, it may be sieved immediately and air-dried. If the sample is too wet to reduce in size, it should be weighed before size reduction and air-dried using an oven that is set 10 to 15°C (18 to 27°F) above room temperature with a maximum oven temperature of 40°C (106°F); under ambient temperature conditions, ambient temperature should be used. In this manner, the moisture in the sample is reduced to an equilibrium condition with the air in the laboratory, and changes in moisture content are minimized during the crushing and grinding operations and even during analysis. After reduction of the gross sample to number 4 or number 8 top size, it is divided and a laboratory sample taken. The laboratory sample is then air-dried and reduced to number 8 top size if necessary. If the total moisture is to be determined (ASTM D-3302), residual moisture is determined by heating at 104 to 110°C (219 to 230°F) for approximately 1 hour.

Air drying removes most of the surface moisture of coal, while a temperature of approximately 107°C (225°F) is needed to remove inherent moisture. At temperatures of approximately 200 to 300°C (392 to 572°F), moisture from the decomposition of organic materials is driven off, but water of hydration requires a considerable amount of energy for expulsion. For example, the water of hydration in clay minerals may require a temperature in excess of 500°C (932°F). However, the issues of decomposition moisture and water of hydration of mineral matter are not usually dealt with in conventional analysis because the temperatures specified in the test methods for moisture determination are well below those needed to remove such moisture.

Usually, the first moisture value to be obtained on a coal sample is the *air-dry loss moisture*. This moisture loss occurs during an attempt to bring the coal sample into equilibrium with the atmosphere in the sample preparation room. The practice of using temperatures above room temperature may accelerate oxidation but shortens the time needed for air drying; hence, temperatures above 40 to 50°C (104 to 122°F) are not recommended for air drying.

In the test method for determination of the equilibrium moisture in coal (ASTM D-1412; ISO 1018), a sample is brought into equilibrium in a partially evacuated desiccator with an atmosphere of 96 to 97% relative humidity at 30°C (86°F). However, as in all methods of determining moisture, precautions must be taken to obtain reliable results from this test method. Over-dried and/or oxidized coal results in low moisture values. To prevent overdrying, the sample should be kept wet before this test is run, and using a dry nitrogen atmosphere can prevent oxidation of the coal during the test. During the test itself, it is important to observe the temperature and time limits for equilibration (as specified in the test method). Furthermore, sudden lowering of the temperature or a sudden surge of air into the desiccator after equilibration may cause condensation of moisture on the coal. In addition, loss of part of the coal sample when a sudden surge of air is allowed into the evacuated desiccator voids the results of the test.

Sample handling should be kept at a minimum during moisture determination, thereby eliminating the potential for loss or gain of moisture during prolonged handling. Heat generated by the crushing and grinding operations used during reduction of the gross sample may be sufficient to cause moisture loss. Alternatively, or in addition, the relative humidity of the sample during preparation and the relative humidity in the testing laboratory may also change during the time that is required for a complete analysis. Air-drying steps in the analysis and efficient sample handling help minimize the effects of relative humidity changes.

Exposure of the coal sample to the atmosphere for extended periods of time increases the chances of oxidation, which results in a weight gain by the coal sample that offsets part of the loss of moisture and gives moisture results that are incorrect. In the determination of moisture by a weight-loss method, it is necessary to attain a constant weight, which requires alternate heating and cooling of samples. Prolonged heating or an excessive number of alternate heating and cooling cycles should be avoided, to minimize the chances of oxidation.

The natural bed moisture of coal is determined (ASTM D-1412; ISO 1018) by wetting the coal, removing the excess water by filtration, and allowing moisture equilibration to occur by standing the coal over a saturated solution of potassium sulfate in a closed vessel, thereby maintaining the relative humidity at 96 to 97%. The vessel must be evacuated to about 30 mm Hg and the entire sample maintained at 30 ± 0.2°C (86 ± 0.4°F) for 48 hours for coals of higher rank than lignite; lignite will require 72 hours to reach equilibrium. The method can also be used to estimate the surface for extraneous moisture of wet coal; such moisture is the difference between the total moisture of the coal and the natural bed moisture.

3.1.2 Data Handling and Interpretation

Moisture values are very important, due to the influence they have on other measured and calculated values used in coal analysis and, ultimately, to the part they play in the buying and selling of coal. The moisture values obtained from the various drying procedures are expressed as a percentage, by weight, of the sample used in the particular test. Consequently, a correction factor must be used to make the various moisture values additive so that total moisture values can be obtained.

The various forms of moisture in coal are described according to the manner in which they are measured by some prescribed standard method. These forms are (1) inherent moisture, (2) surface or free moisture, (3) total moisture, (4) air-dry loss moisture, (5) residual moisture, (6) as-received moisture, (7) decomposition moisture, and (8) water of hydration of mineral matter.

Inherent moisture (*bed moisture, equilibrium moisture, capacity moisture*) is assumed to be the water held within the pore system and capillaries of coal and is not to be identified with *residual moisture. Surface moisture* (*free moisture*) is, as the term implies, water held on the surface of the coal. *Total moisture* is the moisture determined as the loss in weight in an air atmosphere under rigidly controlled conditions of temperature, time, and airflow (ASTM D-3302) and is the sum of *inherent moisture* and *free moisture* and is also the sum of the *air-dry loss* and *residual moisture. Air-dry loss moisture* is the loss in weight resulting from the partial drying of coal, and *residual moisture* is that remaining in the sample after determining the air-dry loss moisture. *As-received moisture* also is equal to the total moisture, or is the sum of the inherent and free moisture present in the coal at the time of the analysis. *Decomposition moisture* is produced from the thermal decomposition of organic constituents of coal. *Water of hydration of mineral matter* is the water that is incorporated into the crystal lattices of the clay and inorganic minerals in coal.

The total moisture is used for calculating other measured quantities to the as-received basis. Total moisture is important in assessing and controlling the commercial processing of coals. It is used to determine the amount of drying that is needed to reach a given moisture requirement and to determine the amount of dust-proofing and freeze-proofing agents to add. In coking processes, coals with a high moisture content require more heat for vaporization of the moisture, which leads to longer coking cycles and decreased production. The total moisture of the coal used must be accurately known to allow for proper charging of the coke ovens and overall control of the coking process.

Inherent or equilibrium moisture is used for calculating moist, mineral-matter-free calorific values for the rank classification of high-volatile bituminous coals. It is also used for estimating free or surface moisture, since total moisture is equal to the sum of the inherent moisture and the free moisture and is considered the inherent moisture of the coal as it occurs in the unexposed seam, where the relative humidity is probably near 100%. However, due to physical limitations, equilibrium moisture determinations are made at 96 to 97% relative humidity and used as inherent moisture values.

Surface moisture is obtained by subtracting equilibrium moisture from total moisture. However, there is no sharp dividing line between inherent moisture and surface moisture. The measurement of inherent moisture depends on the fact that its vapor pressure is less than that of surface moisture. Drying, pulverizing, dust-proofing, and the general handling of coal all depend on surface moisture data since wet coal is very difficult, and in some instances almost impossible, to pulverize.

There is no simple and reliable method of determining the water of hydration of mineral matter. The average value of 8% of the ash is used as the value for water of hydration of mineral matter in coals in the United States. This value is acceptable, although it is an average of values that range from 2 to 3% and up to 15 to 20%. Water of hydration values are used to correct ash to the form of hydrated minerals in mineral matter calculations.

3.2 ASH

Ash is the residue remaining after the combustion of coal under specified conditions (ASTM D-3174; ISO 1171) and is composed primarily of oxides and sulfates. It should not be confused with mineral matter, which is composed of the unaltered inorganic minerals in coal (Given and Yarzab, 1978; Elliott, 1981, and references cited therein; Speight, 1994, and references cited therein). Thus, ash is formed as the result of chemical changes that take place in the mineral matter during the *ashing* process. The quantity of ash can be more than, equal to, or less than the quantity of mineral matter in coal, depending on the nature of the mineral matter and the chemical changes that take place in ashing. The various changes that occur include (1) loss of water from silicate minerals, (2) loss of carbon dioxide from carbonate minerals, (3) oxidation of iron pyrite to iron oxide, and (4) fixation of oxides of sulfur by bases such as calcium and magnesium. In fact, incineration conditions determine the extent to which the weight changes take place, and it is essential that standardized procedures be followed closely to ensure reproducibility.

There are two types of minerals in coal: (1) extraneous mineral matter and (2) inherent mineral matter. *Extraneous mineral matter* consists of materials such as calcium, magnesium, and ferrous carbonates; pyrite; marcasite; clay; shale; sand; and gypsum. *Inherent mineral matter* represents the inorganic elements combined with organic components of coal that originated from the plant materials from which the coal was formed.

The use of coal with mineral matter that gives a high-alkali-oxide ash often results in the occurrence of slagging and fouling problems. As oxides, most ash elements have high melting points, but they tend to form complex compounds (often called *eutectic mixtures*) that have relatively low melting points. On the other hand, high-calcium, low-iron ash coals tend to exhibit a tendency to produce low-melting-range slag, especially if the sodium content of the slag exceeds about 4% w/w.

The chemical composition of coal ash is an important factor in fouling and slagging problems and in the viscosity of coal ash in wet bottom and cyclone furnaces. The potential for the mineral constituents to react with each other (Table 3.2) (Given and Yarzab, 1978) as well as to undergo significant mineralogical changes is high (Helble et al., 1989). In addition, coal with a

TABLE 3.2 Behavior of Minerals at High Temperature[a]

Inorganic Species	Behavior on Heating	Consequences for Analysis
Clays	Lose structural OH groups with rearrangements of structure and release of H_2O	Ash weighs less than MM; yield of water increases apparent organic hydrogen, oxygen, and VM
Carbonates	Decompose with loss of CO_2; residual oxides fix some organic and pyritic S as sulfate	Ash weighs less than MM, but this effect partly neutralized by fixation of S as sulfate; CO_2 from carbonates increases apparent VM, organic carbon, and organic oxygen
Quartz	Possible reaction with iron oxides from pyrite and organically held Ca in lignites; otherwise, no reaction	None, unless reactions indicated take place
Pyrite	In air, burns to Fe_2O_3 and SO_2; in VM test, decomposes to FeS	Increases heat of combustion; ash weighs less than MM; S from FeS_2 contributes to VM
Metal oxides	May react with silicates	None (?)
Metal carboxylates (lignites and subbituminous only)	Decompose, carbon in carboxylate may be retained in residue	Uncertainty about significance of ash; most of organic sulfur in coal fixed as sulfate in ash

[a] MM, mineral matter; VM, volatile matter.

high-iron-content (usually >20% w/w ferric oxide) ash typically exhibits ash-softening temperatures below 1205°C (2200°F).

In some test methods it is recommended that the color of the ash be noted, as it gives an approximate indication of the fusion point. Generally, highly colored ash has a low fusion point, whereas white ash, provided that there are essentially no basic oxides, has a high fusion point.

3.2.1 Test Method Protocols

The test method used for determining the ash content of coal (ASTM D-3174) for proximate and ultimate analysis is not always applicable to the preparation of ash for identification of the ash constituents because of the loss of ash constituents during the ashing procedure.

The determination of mineral ash in coal is usually by heating (burning) an accurately weighed sample of the coal in an adequately ventilated muffle furnace at temperatures in the range 700 to 750°C (1290 to 1382°F) for 4 hours (ASTM D-3174). Typically, the experimental data should be reproducible within ±0.2% of the end result. Other standards (ISO 1171) may vary and require somewhat higher temperatures for the determination of the ash in coal. There are also other methods, predominantly thermal, which can be used for the analysis of coal ash (Voina and Todor, 1978).

The important weights in this determination are those with the coal in the dish before and after drying, the difference between them giving the weight of moisture expelled. These must be determined, therefore, as accurately as possible, and to enable this to be accomplished, the weight of the dish alone should be as small as possible. Suitable petri dishes of thin glass measuring 5 cm in diameter and 1 cm in height meet all the requirements of weight and area.

If the coal under investigation oxidizes appreciably at 100°C (212°F), it is often necessary to carry out the determination in an inert atmosphere either by passing a current of oxygen-free nitrogen through the oven or by using an oven that can be evacuated. In either case it is an advantage to keep the free space in the oven down to a minimum so as to render more effective the displacement action of the nitrogen or the attainment of a suitable vacuum.

Variations in the amount of ash arise from the retention of sulfur that originates from the pyrite. Sulfur in ash is usually determined as sulfate (ASTM D-1757; ASTM D-5016), and the method may give abnormally high amounts of sulfur. This is due to the sulfur retention from pyrite (and marcasite). If the forms of sulfur in coal are known (ASTM D-2492), the amount of pyrite retention can be estimated (see also ASTM D-3174, Note 2). Nevertheless, sulfur retention will give rise to anomalous results.

For high-rank coal, if the amount of pyrite and carbonate minerals is low, sulfur retention is not a major issue, and ashing may be carried out rapidly. For coal that has substantial amounts of pyrite and calcite, the preferred procedure involves burning the coal at low temperatures to decompose the pyrite before the decomposition point of the carbonate minerals is reached, and less sulfur remains in the coal to react with the oxides that are formed at higher temperatures. Another method to combat sulfur retention consists of (1) ashing the coal, (2) treating the ash residue with sulfuric acid, and (3) igniting to a constant yield of ash. The amount of ash is calculated back to the calcium carbonate basis by subtracting three times the equivalent of carbon present as mineral carbonate from the ash as weighed.

In light of the fact that coal ash is measured by removal (combustion) of the organic part of the coal, there are methods for measuring the mineral matter content of coal. Such a method involves demineralization of the coal that depends on the loss of weight of a sample when it is treated with aqueous hydrofluoric acid at 55 to 60°C (130 to 140°F) (Radmacher and Mohrhauer, 1955; Given and Yarzab, 1978; ISO 602). However, pyrite is not dissolved by this treatment and must be determined separately. Other methods include the use of physical techniques such as scanning electron microscopy and x-ray diffraction (Russell and Rimmer, 1979).

A compositional analysis of the ash in coal is often useful in the total description of the quality of the coal. Knowledge of the composition of ash is useful in predicting the behavior of ashes and slags in combustion chambers. The amount and composition of ash is important in determining the best cleaning methods for coals, in selecting coals to be used in the production of coke, and in selecting pulverizing equipment to be used in commercial pulverizing operations. Utilization of the ash by-products of coal combustion sometimes depends on the chemical composition of the ash.

A wide range of trace elements occurs in coal, primarily as a part of the mineral matter. The release of certain trace elements into the environment as combustion products or in the disposal of ash is a concern for coal-burning facilities. Determination of certain trace elements in coal and coal ash is becoming an increasingly important part of coal analysis.

In the preparation of ash samples for analysis, slow burning of the coal samples is necessary to prevent the retention of sulfur as sulfate in the ash. The retention of sulfur as calcium sulfate ($CaSO_4$) in coal ash [900°C (1650°F)] is related to the amounts of calcium and sulfur in whole coal and the ashing temperature to which it is subjected. Iron oxides formed during oxidation of pyrite contributed to the catalytic oxidation of sulfur dioxide (SO_2) to sulfur trioxide (SO_3), which reacts with calcium to form calcium sulfate ($CaSO_4$) during the ashing process. The amount of sulfur retained in power plant ash formed at temperatures exceeding 1000°C (1832°F) is related directly to the concentration of unburned carbon. If the rate of burning is too rapid, some of the sulfur oxides produced from burning pyrite may react with metal oxides to form stable sulfates. The result is that indefinite amounts of sulfur are retained, which introduces an error into all the analytical results unless all other items are corrected to the SO_3-free basis.

In selected test methods (ASTM D-3682; ASTM D-4326), the ash is prepared by placing a weighed amount (usually 3 to 5 g) of coal (sieved to the specified size) in a cold muffle furnace. The temperature is gradually raised to 500°C (932°F) in 1 hour and to 750°C (1382°F) in 2 hours. The analysis of coal ash involves a combination of methods to determine the amounts of silica (SiO_2), alumina (Al_2O_3), ferric oxide (Fe_2O_3), titanium oxide (TiO_2), phosphorus pentoxide (P_2O_5), calcium oxide (CaO), magnesium oxide (MgO), sodium oxide (Na_2O), and potassium oxide (K_2O).

A variety of methods have been recommended for the analysis of coal ash, including: (1) atomic absorption (ASTM D-3682; ASTM D-3683; ASTM D-4606; ASTM D-6414), (2) inductively coupled plasma–atomic emission spectroscopy (ASTM D-6349; ASTM D-6357), and (3) x-ray fluorescence (ASTM D-4326).

Slagging, fouling, and clinkering difficulties have been found to correlate not only with the composition of coal ash but also with the *fusibility* of the coal ash (ASTM D-1857; ISO 540) insofar as ash fusibility is related to composition. The critical temperature most commonly referenced in evaluation of the properties of coal ash is the softening temperature.

In the test for the fusibility of coal ash (ASTM D-1857; ISO 540), the temperatures are observed at which triangular cones prepared from the ash begin to deform and then pass through specified stages of fusion when heated at a specified rate. The test procedure provides for performance of the test in a controlled atmosphere. It is first performed in a mildly reducing atmosphere and is then repeated with a second set of cones in an oxidizing atmosphere. The critical temperature most commonly referenced in the evaluation of the properties of coal is the *softening temperature*, the temperature at which the cone has fused down to a spherical lump in which the height is equal to the width at the base. The significance of performing the test in reducing and oxidizing atmospheres is that most oxides of metals exhibit higher fusion temperatures in their highest state of oxidation. In practical combustion there invariably are zones where the supply of oxygen is depleted or carbon dioxide is reversibly reduced to carbon monoxide in the presence of excess hot carbon. This can produce a reducing atmosphere in a hot zone where ash particles in an incipient state of fusion begin to melt. Slagging, fouling, and clinkering difficulties have been found to correlate with the fusibility of the ash.

3.2.2 Data Handling and Interpretation

Several formulas have been proposed for calculating the amount of mineral matter originally in the coal using data from ashing techniques as the basis of the calculations. Of these formulas, two have survived and have been used regularly to assess the proportion of mineral matter in coal: the Parr formula and the King–Mavies–Crossley formula.

In the Parr formula, the mineral matter content of coal is derived from the expression

$$\text{mineral matter (\% w/w)} = 1.08\text{A} + 0.55\text{S} \tag{3.2}$$

where A is the percentage of ash in the coal and S is the total sulfur in coal.

On the other hand, the King–Maries–Crossley formula is a little more complex and attempts to take into account all of the chemical possibilities that can arise when minerals are present:

$$\begin{aligned}\text{mineral matter (\% w/w)} = {} & 1.09\text{A} + 0.5\text{S}_{\text{pyr}} + 0.8\text{CO}_2 - 1.1\text{SO}_3\text{(ash)} \\ & + \text{SO}_3\text{(coal)} + 0.5\text{Cl}\end{aligned} \tag{3.3}$$

where A is again the percentage of ash in the coal, S_{pyr} the percentage of pyrite sulfur in the coal, CO_2 the percentage of carbon dioxide that originates from the mineral matter in the coal, SO_3 (ash) the percentage of sulfur trioxide in the ash, SO_3 (coal) the percentage of sulfur trioxide in the coal, and Cl the percentage of chlorine in the coal.

The ash value is the analytical value most commonly used for evaluating sampling procedures and is one of the values normally specified in coal contracts. In combustion, high ash content reduces the amount of heat obtainable from a given quantity of coal. High ash content also leads to the problem of handling and disposing of larger amounts of ash produced during combustion. The composition of coal ash is considered in the amount of clinkering and boiler tube slagging that may occur in a boiler. The design of most boilers is such that only coals with a specified range of ash content may be used in the efficient operation of the boiler. The amount of ash in coal used in a coking process is an indication of the amount of ash that will remain in the coke that is made. Coke with a high ash content that is used in a blast furnace requires more fluxing limestone to compensate for the ash and a greater volume of coke to obtain the required amount of usable carbon.

Various processes that result in a reduction of the mineral matter and sulfur content can be employed to clean coal. The ash content of raw coal is often used to select the best cleaning method, and the ash content of the cleaned coal is used to measure the effectiveness of the cleaning process. In the commercial pulverization of coals, the amount and nature of ash is considered carefully before selecting pulverizing equipment or setting up the process.

3.3 VOLATILE MATTER

Volatile matter, as determined by the standard test methods (i.e., ASTM D-3175; ISO 562), is the percentage of volatile products, exclusive of moisture vapor, released during the heating of coal or coke under rigidly controlled conditions. The measured weight loss of the sample corrected for moisture establishes the amount of material (volatile matter) evolved from the coal under the conditions of the test. However, the method, being empirical, requires close adherence to detailed specifications, and since the test is essentially an assay of the sample of coal on a small scale rather than a purely chemical test, it is necessary in order that results may be comparable among laboratories that the conditions prescribed be followed rigidly.

The type of heating equipment and the size and shape of the sample holders as well as the material from which they are made (platinum crucibles specified) all have some influence on the rate of heating of the sample and the range of temperatures to which it is exposed. The crucibles used are 10 to 20 mL in capacity, of specified size, with deep-fitting lids, and there are two procedures for determination of volatile matter.

3.3.1 Test Method Protocols

Determination of the volatile matter content of coal (ASTM D-3175; ISO 562) is an important determination because volatile matter data are an integral part of coal classification systems (Chapter 1) and form the basis of evaluating coals for their suitability for combustion and carbonization. The methods for determining volatile matter content are based on the same principle and consist of heating a weighed sample of coal (usually about 1 g) in a covered crucible to a predetermined temperature; the loss in weight (excluding losses due to water) is the volatile matter content (expressed as a weight percent).

In the test method (ASTM D-3175), 1 g of coal is weighed and placed in a preweighed platinum crucible (10 to 20 mL in capacity, 25 to 35 mm in diameter, and 30 to 35 mm in height) with a close-fitting cover. The crucible is then suspended at a specified height in the furnace chamber. The temperature of the region in the furnace where the crucible is suspended must be maintained at $950 \pm 20°C$ ($1742 \pm 36°F$). After the more rapid discharge of volatile matter, as evidenced by the disappearance of the luminous flame, the cover of the crucible should be tapped to ensure that the lid is still properly seated to guard against the admission of air. After heating for exactly 7 minutes, the crucible is removed from the furnace and cooled. The crucible should be weighed as soon as it is cold. The percentage loss of weight minus the percentage moisture equals the volatile matter. The residue in the crucible is coke, and it is often advantageous to indicate the nature of the coke button obtained, since it provides a useful guide to the caking properties of the coal. It is usually described under the following headings: color, luster, swelling, fissuring, structure, both internal and surface, and hardness.

If decrepitation or entrainment takes place, the coal sample may be either heated more slowly or blended with a coking coal (the total volatile matter of which is known) in the ratio 4:1 and the test carried out with the blend.

A modified procedure is used for coal that does not yield a coherent cake as residue in the determination and that evolve gaseous products at a rate sufficient to carry solid particles out of the crucible when heated at the standard rate. Such coal, referred to as *sparking coal*, normally includes all low-rank noncaking coal and lignite but may include other coal as well (ASTM D-121; ASTM D-3175). In this procedure (ASTM D-3175, Sec. 7.3), the sample is suspended and heated in a cooler zone of the furnace such that the temperature inside the crucible reaches $600 \pm 50°C$ ($1112 \pm 90°F$) in 6 minutes. After the preliminary heating, the crucible is lowered into the hot zone ($950 \pm 20°C$, $1742 \pm 36°F$) of the furnace and held there for 6 minutes. The crucible is then removed from the furnace and set on a metal block to cool before weighing. The cooling period should be kept constant and should not exceed 15 minutes to ensure uniformity of results. The volatile matter is calculated in the same manner as in the regular method.

The rate of heating of the sample influences volatile matter values and makes it necessary to calibrate equipment to achieve a satisfactory and reproducible heating rate. This calibration can be accomplished by using either a manual or an automatic mechanical device that lowers the sample crucible into the electrically heated furnace at a reproducible rate.

The crucibles and covers must be properly shaped to ensure a proper fit. A loose-fitting cover allows air to come in contact with the hot coal sample, with subsequent formation of volatile carbon oxides. Such an occurrence would result in a volatile matter value that is too high. Oxidation is not a serious problem in volatile matter determinations, since the rapid release of large amounts of gases during the test does prevent the entry of air into the crucible, thereby reducing the chance of oxidation. Addition of a few drops of a volatile material, such as toluene, may also help prevent oxidation.

The standard crucible method used for coal may be unsuitable for coke because coke often requires a higher temperature for removal of the last remnants of the volatile matter. A thermogravimetric procedure in nitrogen at 1000°C (1830°F), or higher if necessary, for a specified period (30 minutes or more) may be desirable for coke. However, the precise conditions must be acknowledged formally when the yield of volatile matter is declared.

The specification giving maximum clearance between the crucible and the lid is to standardize the amount of oxygen admitted to the heated sample. Unlimited oxygen results in complete conversion of the organic material to carbon dioxide and water. Limited oxygen provides for the formation of coke. This is analogous to the carbon residue tests so often used in petroleum technology (Speight, 2001, 2002).

The composition of the volatile matter evolved from coal is, of course, substantially different for the different ranks of coal, and the proportion of incombustible gases increases as the coal rank decreases. Furthermore, in macerals isolated from any one particular coal, the volatile matter content decreases in a specific order; thus, exinite produces more volatile matter than vitrinite, which, in turn, yields more volatile matter than inertinite.

The arbitrary nature of the test for volatile matter content precludes a detailed discussion of the various national standards. However, in general terms, the temperature is in the range 875 to 1050°C (1605 to 1920°F), the duration of heating is 3 to 20 minutes, and the crucibles may be platinum, silica, or ceramic material. The German standard specifies a temperature of 875°C (1605°F) to be in accord with industrial coking practice; other standards specify temperatures of 1000 to 1050°C (1830 to 1920°F) to ensure maximum evolution of volatile matter under the test conditions: a temperature of 950 ± 20°C (1740 ± 35°F) is specified (ASTM D-3175).

The various standards usually specify that a single crucible be employed, although there has been a tendency in France and Belgium to advocate the use of two crucibles. In this method, the coal is heated in a crucible that is enclosed in a larger crucible, with the space between the two crucibles filled with carbon (charcoal).

The chief differences in the methods for the determination of volatile matter emanating from the thermal decomposition of coal are (1) variations in the size, weight, and materials of the crucibles used; (2) the rate of temperature rise; (3) the final temperature; (4) the duration of heating; and (5) any modifications that are required for coals which are known to decrepitate or which may lose particles as a result of the sudden release of moisture or other volatile materials. In essence, all of these variables are capable of markedly affecting the result of the tests, and it is, therefore, very necessary that the standard procedures be followed closely.

The advantage of using two crucibles in methods of this type is believed to arise because of the need to prevent partial oxidation and hence the reduction of accrued errors due to loss of material as carbon oxides. Indeed, the very nature of the test for volatile matter content (which is not a determination of low-molecular-weight volatile matter in the coal but is more a test for volatile matter formed as a result of a wide variety of decomposition reactions during the test) dictates that partial oxidation reactions be excluded.

Furthermore, the lower rate of heating (resulting from the insulating effect of the charcoal) assists in preventing the election of solid particles, which can occur when lower rank coals (in 10- to 20-mL platinum crucibles) are heated at the higher rate by direct insertion into the hot furnace. There is, therefore, no need for a modification of the method when it is applied to the lower-rank coals (an advantage), since in borderline cases, the result may be up to 2% lower (i.e., 32% volatile matter instead of 34% volatile matter) when the modification (the lower heating rate) is employed.

Mineral matter may also contribute to the volatile matter by virtue of the loss of water from the clays, the loss of carbon dioxide from carbonate minerals, the loss of sulfur from pyrite (FeS_2), and the generation of hydrogen chloride from chloride minerals as well as various reactions that occur within the minerals, thereby influencing the analytical data (Given and Yarzab, 1978).

The characterization of coal either as agglomerating or as nonagglomerating for the purposes of rank classification is carried out in conjunction with the determination of the volatile matter content. Thus, if the residue remaining from the determination is in the form of an agglomerate button that is capable of supporting a 500-g weight without pulverization of the button or if the button shows swelling or cell structure, the coal is classified as agglomerating.

3.3.2 Data Handling and Interpretation

It is actually inaccurate to describe a coal as *containing 35% volatile matter* or as a coal *with 35% volatile matter* since the volatile matter is not present as such in the coal. The volatile matter is simply those products of the thermal decomposition of the coal substance that are expelled from the standard crucible under the standard conditions of the test, and consequently, a coal should be described accurately as *yielding 35% volatile matter*. Although the first two statements are those most widely used and generally accepted in fuel technology literature, it should always be remembered that the suggestions they imply are incorrect.

Determination of volatile matter content using a slower heating rate is applicable to a wider variety of coals. However, the values obtained are sometimes lower (1 to 3% absolute) than those obtained from the regular method. This illustrates the empirical nature of this test and the importance of strict adherence to detailed specifications. The complexity of the constituents of coal that undergo decomposition during this test makes it necessary to have wide tolerances for reproducibility and repeatability.

Volatile matter values are important in choosing the best match between a specific type of coal-burning equipment and the coal to use with the equipment. Such values are valuable to fuel engineers in setting up and maintaining proper burning rates. Volatile matter values are used as an indication of the amount of smoke that may be emitted from furnaces or other types of coal-burning equipment. Limits may be set on the volatile matter content of the coal used in certain coal-burning facilities to control smoke emissions.

In addition, the quantity of volatile matter released by coal during slow heating (beginning at room temperature) differed from the amounts released by rapid heating (i.e., the instant application of final temperature). When the coal is heated slowly, the volatile matter yield decreases and fixed carbon increases. In addition, there is a relationship between the yield of volatile matter released by coal during rapid pyrolysis and the temperature to which the coal is subjected as well as to the atomic hydrogen/carbon ratio of the coal. Over the range 200 to 1000°C (390 to 1830°F) the relationship is linear, and partially treated coal, in which only part of the volatile matter is driven off, when heated to a higher temperature yields the calculated additional amount of volatile matter.

3.4 FIXED CARBON

Fixed carbon is the material remaining after the determination of moisture, volatile matter, and ash. It is, in fact, a measure of the solid combustible material in coal after the expulsion of volatile matter, and like determination of the carbon residue of petroleum and petroleum products (Speight, 1999, 2001) represents the approximate yield of thermal coke from coal (Zimmerman, 1979).

The fixed-carbon value is one of the values used in determining the efficiency of coal-burning equipment. It is a measure of the solid combustible material that remains after the volatile matter in coal has been removed. For this reason, it is also used as an indication of the yield of coke in a coking process. Fixed carbon plus ash essentially represents the yield of coke. Fixed-carbon values, corrected to a dry, mineral-matter-free basis, are used as parameters in the coal classification system (ASTM D-388).

Data reporting (i.e., the statement of the results of the proximate analysis test methods) usually includes (in some countries but not in all countries) descriptions of the color of the ash and of the coke button. As an interesting comparison, the test for determining the carbon residue (Conradson), the coke-forming propensity of petroleum fractions and petroleum products (ASTM D-189; ASTM D-2416), advocates the use of more than one crucible. A porcelain crucible is used to contain the sample, and this is contained within two outer iron crucibles. This corresponds to the thermal decomposition of the sample in a limited supply of air (oxygen) and the measurement of the carbonaceous residue left at the termination of the test.

3.5 COAL ASSAY

In some instances the information gained from the proximate analysis coal sample requires being supplemented by data that provides a more detailed examination of the behavior of coal during carbonization. Such a test provides data relating to the yields and qualities of coke, tar, liquor, and gas that can be correlated with those obtained in large-scale practice. Two types of assay have been developed, one for use at 600°C (1112°F) that can be related to low-temperature carbonization processes, and a second at 900°C (1652°F) that gives results allied to those obtained in high-temperature carbonization practice such as gas and coke manufacture.

3.5.1 Gray–King Assay at 600°C (1112°F)

In the test method the charge of coal is heated in a silica retort by means of an electric furnace that is arranged so that it can be pushed over the retort after having been heated to the necessary temperature. The volatile products pass from the retort to a condenser that can be cooled, if necessary, in a vessel containing a cooling medium and in which the tar and liquor collect. The remaining gases are freed from ammonia by passage through a glass scrubber containing glass beads drenched with dilute sulfuric acid and then pass into the gas holder, a (3- to 4-L) vessel containing water saturated with gas from previous assays, where they are collected at constant pressure by means of the arrangement; any evolved water is collected in a glass or metal jar.

During the first period of heating, observe the temperatures at which the following events occur: (1) appearance of oil vapor, (2) evolution of gas as distinct from displacement due to expansion, and (3) further evolution of gas, at the onset of which the rate of discharge of water increases suddenly. These three temperatures are usually clearly defined and are characteristic of the sample being assayed.

Correlation of the data with industrial practice is easily accomplished and the behavior of any coal on carbonization on a large scale is fairly accurately predictable. For example, (1) the coke yield and volatile matter remaining in the coke are approximately equal to those furnished on a large scale; (2) the gas yield is usually slightly greater than the large-scale yield, mainly because of the necessity in industrial practice to reduce the period of carbonization as much as possible; (3) the tar yield is much greater than that given by a large-scale technique (normally, the commercial yield varies from 50 to 80% of the assay yield of tar); (4) the yields of ammonia are comparable; and (5) if an analysis of the gas is required it is advisable to carry out two assays and use the gas from the first to sweep the air out of the assembled apparatus before the second test is begun. The gas obtained from the second experiment will then be practically free of contamination with air.

3.5.2 Gray–King Assay at 900°C (1652°F)

In this assay, the tar vapors are cracked over crushed silica brick and the increased yield of gas is accommodated in a larger receiver of some 10 L capacity. Owing to the possibility of tar fog remaining uncondensed, the condenser in this case is followed by a U-tube packed with absorbent cotton wool, and both are immersed in a mixture of crushed ice and salt. The cracking material is contained in a retort that is heated by two electric furnaces, one carbonizing the 6-in. coal layer at a final temperature of 900°C (1650°F) and the other maintaining the cracking zone at 800°C (1470°F).

Since the high-temperature assay is used mainly for the examination of gas-making coals, the range of coke types normally encountered is considerably smaller than that met with in low-temperature assays, and as a consequence, no standard classification of coke types has been drawn up, nor is such a classification so necessary in this case. Nevertheless, a full description of the coke should always be made.

During the performance of the test, the temperature of the cracking furnace is fairly critical, especially during the early stages of the carbonization. With the carbonization furnace increasing in temperature at the prescribed rate, the influence of this on the temperature of the cracking furnace is appreciable when it is remembered that the cracking furnace current had originally been adjusted to maintain 800°C (1470°F) against room temperature.

Strongly swelling coals should be carbonized either in admixture with electrode carbon or, preferably, without dilution, in a special retort in which the section containing the coal has been enlarged in diameter to 30 mm, the remaining part of the tube being unaltered. This permits a 20-g charge of a highly swelling coal to be carbonized without the necessity of adding electrode carbon.

3.5.3 Other Carbonization Tests

In those circumstances in which the detailed information furnished by the Gray–King assay is not essential, one or other of two simpler tests that, in practice, are frequently carried out in addition to the assay may be used to assess the caking characteristics of coal. These are the *agglutinating value test* and the *free-swelling index test*. The former gives a measure of the caking properties of a coal; the latter indicates the degree of swelling that a sample of coal will undergo during carbonization, both factors of importance in the carbonization industries.

In the *agglutinating value test*, the caking properties of the coal are progressively destroyed by the addition of sand of specified characteristics until the coal–sand mixture, on carbonization under controlled conditions, just fails to satisfy one or both of two conditions.

The *agglutinating value* (sometimes referred to as the *caking index*) of coal is the maximum whole-number ratio of sand to coal in a 25-g mixture which, on carbonization at 900 ± 15°C (1650 ± 27°F) for 7 minutes in a standard silica crucible, produces a carbonized mass capable of supporting a 500-g weight and yielding less than 5% of loose, unbound material. It is essential for this test that not only are the experimental conditions rigidly observed but the coal must always be ground to the same degree of fineness to ensure strictly comparable figures (different size distributions may lead to widely different agglutinating values) and the sand used must be of uniform quality in respect to sharpness and purity as well as particle size.

An approximate idea of the degree of swelling (the *swelling index*) experienced by a coal on carbonization may be obtained from the coke button left after the determination of volatile matter, but in many cases this measure of the swelling power is not sufficiently discriminating, principally because this is not the main function of the test.

For this reason, a standard crucible test has been devised in which the chief features, compared with other tests, are the use of a leveled coal surface, a crucible of special shape and unidirectional heating by a gas flame. As a result, regularly shaped coke buttons are obtained and the degree of swelling is assessed by comparison with profiles of standard buttons.

The conditions for obtaining the correct heating should be checked by means of a fine wire thermocouple inserted through a pierced lid having its unprotected junction in contact with the center of the base of the empty crucible. The end of the couple should be formed into a flattened loop so that the junction and a portion of each wire rests on the bottom of the crucible during a temperature measurement. The couple should be made of wires not heavier than 34-gauge platinum or 26-gauge base metal.

As an extension of the proximate analysis or coal assay, it is worthy of note that new methods continued to be developed. For example, thermogravimetric analysis has been extended to cover determinations of volatile matter (as well as determination of moisture and ash) in coal and coke. These constituents can be measured by pyrolyzing the samples in oxygen and air, and the weight loss at prescribed temperatures was measured by using a thermobalance.

Fixed carbon and percentages of moisture, volatile matter, and ash were measured in coal in less than 15 minutes using a thermogravimetric analyzer. For coal analysis, the sample is purged with nitrogen at 100°C (212°F) to dry the sample and remove oxygen. The percentage of moisture (weight loss) is read at this point. The loss of volatile matter can be measured by heating the sample to 600°C (1112°F) at 160°C/min (288°F/min) and heating at 600°C (1112°F) for 2 minutes; switching the purge gas to oxygen while heating to 900°C (1652°F) yields a residual weight equivalent to the ash content. Results from tests using this method have not yet been compared with results from tests using standard coal procedures.

3.6 OTHER METHODS

Although in many laboratories the methods described above remain the methods of choice for determining the proximate analysis of coal, there is also a test method for the proximate analysis of coal by instrumental procedures, assuming that calibration is an integral part of the procedure (ASTM D-5142). This method covers the determination of moisture, volatile matter, and ash and the calculation of fixed carbon in the analysis of coal and coke samples prepared in accordance with standard protocols (ASTM D-2013). The results may require a correction for bias or be corrected for instrument calibration using samples of known proximate analysis. However, the standard test methods (ASTM D-3173; ASTM D-3174; ASTM D-3175) are considered to be the referee test methods. The instrumental test methods are not applicable to thermogravimetric analyzers using microgram-size samples. The moisture value determined can be used for calculating other analytical results to a dry basis, and when used in conjunction with the air-dry moisture loss (ASTM D-2013; ASTM D-3302), the calculated total moisture can be used to convert dry basis analytical results to the as-received basis. In addition, the result of the ash determination can be applied in ultimate analysis (ASTM D-3176) for converting the analytical data to an ash-free basis or for application in the equations related to mineral matter content.

Briefly, and as expected, moisture is determined by establishing the loss in mass of the analysis specimen of coal or coke when heated under rigidly controlled conditions of temperature, time, atmosphere, sample mass, and equipment specifications. Volatile matter is determined by measuring the loss in mass of the moist or dried analysis specimen of coal or coke when heated under rigidly controlled conditions. The weight loss measured is the yield of volatile matter after correcting for the moisture content. Ash is determined by the amount of residue after burning the coal under rigidly controlled conditions of sample weight, temperature, time, atmosphere, and equipment specifications. In these instrumental methods, moisture, volatile matter, and ash may be determined sequentially in a single instrumental procedure.

REFERENCES

Allardice, D. J., and Evans, D. G. 1978. In *Analytical Methods for Coal and Coal Products*, Vol. I, C. Karr, Jr. (Editor). Academic Press, San Diego, CA, Chap. 7.

ASTM. 2004. *Annual Book of ASTM Standards*, Volume 05.06. American Society for Testing and Materials, West Conshohocken, PA. Specifically:

ASTM D-121. Standard Terminology of Coal and Coke.

ASTM D-189. Standard Test Method for Conradson Carbon Residue of Petroleum Products.

ASTM D-388. Standard Classification of Coals by Rank.

ASTM D-720. Standard Test Method for Free-Swelling Index of Coal.

ASTM D-1412. Standard Test Method for Equilibrium Moisture of Coal at 96 to 97 Percent Relative Humidity and 30°C.

ASTM D-1757. Standard Test Method for Sulfate Sulfur in Ash from Coal and Coke.

ASTM D-1857. Standard Test Method for Fusibility of Coal and Coke Ash.

ASTM D-2013. Standard Practice of Preparing Coal Samples for Analysis.

ASTM D-2416. Standard Test Method for Coking Value of Tar and Pitch (Modified Conradson).

ASTM D-2492. Standard Test Method for Forms of Sulfur in Coal.

ASTM D-2961. Standard Test Method for Single-Stage Total Moisture Less Than 15% in Coal Reduced to 2.36-mm (No. 8 Sieve) Topsize.

ASTM D-3172. Standard Practice for Proximate Analysis of Coal and Coke.

ASTM D-3173. Standard Test Method for Moisture in the Analysis Sample of Coal and Coke.

ASTM D-3174. Standard Test Method for Ash in the Analysis Sample of Coal and Coke from Coal.

ASTM D-3175. Standard Test Method for Volatile Matter in the Analysis Sample of Coal and Coke.

ASTM D-3176. Standard Practice for Ultimate Analysis of Coal and Coke.

ASTM D-3302. Standard Test Method for Total Moisture in Coal.

ASTM D-3682. Standard Test Method for Major and Minor Elements in Combustion Residues from Coal Utilization Processes.

ASTM D-3683. Standard Test Method for Trace Elements in Coal and Coke Ash by Atomic Absorption.

ASTM D-4326. Standard Test Method for Major and Minor Elements in Coal and Coke Ash by X-ray Fluorescence.

ASTM D-4606. Standard Test Method for Determination of Arsenic and Selenium in Coal by the Hydride Generation/Atomic Absorption Method.

ASTM D-5016. Standard Test Method for Sulfur in Ash from Coal, Coke, and Residues from Coal Combustion Using High-Temperature Tube Furnace Combustion Method with Infrared Absorption.

ASTM D-5142. Standard Test Methods for Proximate Analysis of the Analysis Sample of Coal and Coke by Instrumental Procedures.

ASTM D-6349. Standard Test Method for Determination of Major and Minor Elements in Coal, Coke, and Solid Residues from Combustion of Coal and Coke by Inductively Coupled Plasma–Atomic Emission Spectrometry.

ASTM D-6357. Test Methods for Determination of Trace Elements in Coal, Coke, and Combustion Residues from Coal Utilization Processes by Inductively Coupled Plasma Atomic Emission, Inductively Coupled Plasma Mass, and Graphite Furnace Atomic Absorption Spectrometries.

ASTM D-6414. Standard Test Method for Total Mercury in Coal and Coal Combustion Residues by Acid Extraction or Wet Oxidation/Cold Vapor Atomic Absorption.

BS. 2003. *Methods for Analysis and Testing of Coal and Coke*. BS 1016. British Standards Association, London.

Elliott, M. A. (Editor). 1981. *Chemistry of Coal Utilization*. Wiley, Hoboken, NJ.

Given, P. H., and Yarzab, R. F. 1978. In *Analytical Methods for Coal and Coal Products*, Vol. 2, C. Karr, Jr. (Editor). Academic Press, San Diego, CA, Chap. 20.

Hampel, M. 1974. *Glueckauf*, 110:129.

Helble, J. J., Srinivasachar, S., Boni, A. A., Kang, S-G., Sarofim, A. F., Beer, J. M., Gallagher, N., Bool, L., Peterson, T. W., Wendt, J. O. L., Shah, N., Huggins, F. E., and Huffman, G. P. 1989. *Proceedings of the Sixth Annual International Pittsburgh Coal Conference*. University of Pittsburgh. Vol. 1. p. 81.

ISO. 2003. *Standard Test Methods for Coal Analysis*. International Organization for Standardization, Geneva, Switzerland.

ISO 331. Determination of Moisture in the Analysis of Coal.

ISO 540. Determination of the Fusibility of Ash: High-Temperature Tube Method.

ISO 562. Determination of Volatile Matter in Hard Coal and Coke.

ISO 589. Determination of the Total Moisture of Hard Coal.

ISO 602. Determination of Mineral Matter.

ISO 687. Determination of Moisture in the Analysis Sample of Coke.

ISO 1015. Determination of the Moisture Content of Brown Coals and Lignites.

ISO 1018. Determination of the Moisture Holding Capacity of Hard Coal.

ISO 1171. Determination of Ash.

ISO 8264. Determination of the Swelling Properties of Hard Coal Using a Dilatometer.

ISO 11722. Hard Coal: Determination of Moisture in the General Analysis Test Sample by Drying in Nitrogen.

Russell, S. J., and Rimmer, S. M. 1979. In *Analytical Methods for Coal and Coal Products*, Vol. 3, C. Karr, Jr. (Editor). Academic Press, San Diego, CA, Chap. 42.

Speight, J. G. 1994. *The Chemistry and Technology of Coal*, 2nd ed. Marcel Dekker, New York.

Speight, J. G. 2001. *Handbook of Petroleum Analysis*. Wiley, Hoboken, NJ.

Speight, J. G. 2002. *Handbook of Petroleum Product Analysis*. Wiley, Hoboken, NJ.

Vahrman, M. 1970. *Fuel*, 49:5.

Voina, N. I., and Todor, D. N. 1978. In *Analytical Methods for Coal and Coal Products*, Vol. 2, C. Karr, Jr. (Editor). Academic Press, San Diego, CA, Chap. 37.

Zimmerman, R. E. 1979 *Evaluating and Testing the Coking Properties of Coal*. Miller Freeman, San Francisco, CA.

4 Ultimate Analysis

The ultimate analysis of coal involves determination of the weight percent carbon as well as sulfur, nitrogen, and oxygen (usually estimated by difference). Trace elements that occur in coal are often included as a part of the ultimate analysis.

The carbon determination includes carbon present as organic carbon occurring in the coal substance and any carbon present as mineral carbonate. The hydrogen determination includes hydrogen present in the organic materials as well as hydrogen in all of the water associated with the coal (Chapter 3). In the absence of evidence to the contrary, all of the nitrogen is assumed to occur within the organic matrix of coal. On the other hand, sulfur occurs in three forms in coal: (1) as organic sulfur compounds; (2) as inorganic sulfides that are, for the most part, primarily the iron sulfides pyrite and marcasite (FeS_2); and (3) as inorganic sulfates (e.g., Na_2SO_4, $CaSO_4$). The sulfur value presented for ultimate analysis may include, depending on the coal and any prior methods of coal cleaning, inorganic sulfur and organic sulfur.

Moisture and ash (Chapter 3) are not determined as a part of the data presented for ultimate analysis but must be determined so that the analytical values obtained can be converted to comparable bases other than that of the analysis sample. In other words, analytical values may need to be converted to an *as-received basis*, a *dry basis*, or a *dry, ash-free basis*. When suitable corrections are made for any carbon, hydrogen, and sulfur derived from the inorganic material, and for conversion of ash to mineral matter, the ultimate analysis represents the elemental composition of the organic material in coal in terms of carbon, hydrogen, nitrogen, sulfur, and oxygen.

The standard method for the ultimate analysis of coal and coke (ASTM D-3176) includes the determination of elemental carbon, hydrogen, sulfur, and nitrogen, together with the ash in the material as a whole. Oxygen is usually calculated by difference. The test methods recommended for elemental analysis include the determination of carbon and hydrogen (ASTM D-3178), nitrogen (ASTM D-3179), and sulfur (ASTM D-3177; ISO 334; ISO 351), with associated determination of moisture (ASTM D-3173) and ash (ASTM D-3174) to convert the data to a moisture-ash-free basis.

Handbook of Coal Analysis, by James G. Speight
ISBN 0-471-52273-2

4.1 CARBON AND HYDROGEN

Carbon and hydrogen, which, respectively, account for 70 to 95% and 2 to 6% by weight (dry, ash-free) of the organic substance of coal, are thought by some to be the most important constituents of coal. Almost all of the carbon and hydrogen in coal occurs in combined form in the complex organic compounds that make up coal. But carbon also occurs in the mineral carbonates, with calcite being the principal component, and hydrogen is also present in the various forms of moisture found in coal.

4.1.1 Test Method Protocols

All methods of determining the carbon and hydrogen content of coal are very similar insofar as elemental carbon and hydrogen is determined by combustion in a closed system (combustion train), and the products of combustion are collected in an absorption train. The percentages by weight of carbon and hydrogen are calculated from the gain in weight of the relevant segments of the absorption train. This method gives the total percentages of carbon and hydrogen in the coal, including any carbon in the carbonates and any hydrogen in free moisture and in water of hydration of silicates. Reporting hydrogen and oxygen in the free water (moisture) associated with the sample may optionally be included or excluded from the hydrogen and oxygen values. The test method recommends making a notation of the option (*as received*, or *dry*, or *dry, ash-free*) used as the basis for reporting results. Hydrogen in the water of hydration of silicates is generally ignored but can introduce errors for coals that contain high amounts of mineral matter. Overall, it is preferable, therefore, that the hydrogen and oxygen associated with the free moisture not be included in the hydrogen and oxygen values and that moisture be reported as a separate item.

Several automated elemental analysis systems suitable for coal analyses allow simultaneous determination of carbon, hydrogen, and nitrogen for multiple samples. In fact, rather than by difference (which also accumulates all the errors of other determinations), oxygen can be determined using a separate combustion system. The systems convert coal (catalytically) to nitrogen, carbon dioxide, and water, which are separated with a gas chromatographic column and detected through a sensor, such as a dual thermal conductivity cell. Sulfur can be determined using a similar system that excludes the oxygen determination.

In practice, the older well-used methods (ASTM D-3178; ISO 609; ISO 625) for determining carbon and hydrogen involve combustion of an exact amount of the coal in a closed system and the products of the combustion (carbon dioxide and water) determined by absorption. The combustion is usually accomplished by placing the ground coal (to pass through, for example, a 60-mesh/250-μm sieve) in a stream of dry oxygen at temperatures on the order of 850 to 900°C (1560 to 1650°F) (ASTM D-3178). Complete conversion of the combustion gases to carbon dioxide and water can be achieved by passing the gases through heated cupric oxide.

As noted above, the equipment used for determination of the carbon and hydrogen content of coals consists basically of two parts: a combustion unit and an absorption train (ASTM D-3178). The combustion unit is made up of three separate electrically heated furnaces and the combustion tube. The first furnace, at the inlet end of the tube, is approximately 130 mm long and can be moved along the tube. The second and third furnaces are 330 mm and 230 mm in length, respectively, and are mounted in a fixed position around the combustion tube. The required operating temperatures are 850 to 900°C (1562 to 1652°F), 850 ± 20°C (1562 ± 36°F), and 500 ± 20°C (932 ± 36°F) for the first, second, and third furnaces, respectively. The combustion tube can be constructed from fused quartz or high-silica glass. The combustion tube is packed with lead chromate ($PbCrO_3$) or silver (Ag) gauze under the third furnace and cupric oxide (CuO) under the second furnace. Oxidized copper gauze plugs are used to contain these components in the tube.

The absorption train is composed of a tube packed with water absorbent, a second tube packed with a carbon dioxide absorbent, and a guard tube packed with equal volumes of the water and carbon dioxide absorbents. Anhydrous magnesium perchlorate ($MgClO_4$) is commonly used for the water absorbent, while sodium hydroxide (NaOH) or potassium hydroxide (KOH) impregnated in an inert carrier is used as the carbon dioxide absorber.

In the analytical procedure, a weighed amount of coal is placed in either a boat (glazed porcelain, fused silica, or platinum) and inserted into the combustion tube under the first furnace, where the sample is burned in oxygen. The combustion products are allowed to flow over the heated copper oxide and lead chromate or silver and into the absorption train. The copper oxide ensures complete combustion of the carbon and hydrogen in the coal, whereas the lead chromate absorbs the oxides of sulfur. If silver gauze is used, both the sulfur oxides and chlorine will be absorbed. The preweighed absorbers in the absorption train absorb water and carbon dioxide, and the percent of carbon and hydrogen in the sample is calculated from the gain in weight of absorbers.

This particular method of combustion analysis for carbon and hydrogen (Liebig method) is used internationally, although some modifications may have been made by the various national standards organizations. The sample size can vary from as little as 1 to 3 mg (microanalysis) to 100 to 500 mg (macroanalysis) with combustion temperatures as high as 1300°C (2370°F). However, the method must ensure complete conversion of all of the carbon in the coal to carbon dioxide and all of the hydrogen to water. Oxides of sulfur and chlorine are removed by passing the products of combustion over heated lead chromate and silver gauze. The carbon (as carbon dioxide) and hydrogen (as water) are calculated from the increase in weight of the absorbents used to collect the water and carbon dioxide. Oxides of sulfur (and chlorine, which may be released in significant amounts) are usually removed from the combustion gases by passage over silver at about 600°C (1110°F) while nitrogen dioxide is removed by lead chromate or manganese dioxide. The preliminary and final procedure involves the passage of dry air (also free of carbon dioxide) through the system in order that the absorption train may be weighed full of air before and after the determination.

It is essential that both air and oxygen be pure. If not, purification can be assured by passage through a purification train consisting of (1) 30% aqueous solution of potassium hydroxide, (2) concentrated sulfuric acid, and (3) two U-tubes charged with soda lime (or other dry absorbent for carbon dioxide) and calcium chloride. It may be advisable to use two purification trains (one for each gas) leading to a common tube for delivery to the combustion train.

Usually, the coal burns smoothly in the stream of oxygen. However, if the combustion unit used for burning the sample is heated too rapidly, volatile matter may be released at such a rate that some of it may pass through the entire system and not be completely burned or absorbed. To prevent this from happening, the temperature of the combustion unit must be at the proper level, and enough time must be allowed for complete combustion. In addition, a proper flow of oxygen must be maintained through the system.

The formation of oxides of nitrogen during the combustion process may lead to slightly high results for carbon and hydrogen, since the oxides are acidic in nature and would be absorbed in the absorption train. There are indications that the hydrogen value would not be greatly influenced but that the carbon value would be higher by an amount depending on the nitrogen content and the nitrogen oxides formed during the combustion. For more precise results, the oxides of nitrogen can be removed by absorption on manganese dioxide, or in some cases lead dioxide, before absorption of the water and carbon dioxide. However, caution is advised in treating any potential error lightly.

Since inorganic carbonates and nitrogen oxides contribute to the carbon value in coal as it is normally determined, consideration of this issue is necessary. Hydrogen values also are usually high, due to the inclusion of the various forms of moisture that are present in coal. All of these factors limit the reliability of carbon and hydrogen data for predicting the amount of combustible carbon and hydrogen in coal.

Finally, and most important, blank determinations should be carried out at regular intervals.

4.1.2 Data Handling and Interpretation

If the coal is not dried, the water collected in the adsorption train also contains water present originally as moisture in the coal (Chapter 3). The hydrogen equivalent to the percentage of moisture in the coal is deducted from the percentage of hydrogen given by the method described above to arrive at the corrected figure, or alternatively, the quantity of moisture present in the weight of coal used is subtracted from the weight collected in the calcium chloride tube before calculating the equivalent weight of hydrogen burned.

Subtracting one-ninth of the determined moisture from the determined hydrogen can make a reasonable correction to the hydrogen value for the moisture in coal. A correction to the hydrogen value for the water of hydration of mineral matter is more difficult. The water of hydration of mineral matter for some

(U.S.) coals has been estimated to be 8% of the ash value. A correction of the hydrogen value for the water of hydration can thus be estimated by multiplying the ash value by 0.08, and one-ninth of this figure will give the correction to be subtracted from the hydrogen determined. Upon making these corrections for the forms of moisture, the value for the hydrogen can be calculated:

$$H = H_{\text{as-determined}} - \frac{2.02}{18.02\left(M_{\text{as-determined}} + \frac{8}{100} \times A_{\text{as-determined}}\right)} \tag{4.1}$$

where H is hydrogen, M moisture, and A ash.

The results of the carbon and hydrogen analysis may be reported on any number of bases, differing from each other in the manner by which moisture values are treated. Inclusion of the hydrogen of moisture and water of hydration of mineral matter in the hydrogen value is common practice for the as-determined and as-received bases. Hydrogen values on a dry coal basis, however, are commonly corrected for the hydrogen of moisture. No corrections are normally made to the determined hydrogen value for the water of hydration of mineral matter, due to the uncertainty of the estimate of its value.

4.2 NITROGEN

Nitrogen occurs almost exclusively in the organic matter of coal. Very little information is available concerning the nitrogen-containing compounds present in coal, but they do appear to be stable and are thought to be primarily heterocyclic. The original source of nitrogen in coal may have been both plant and animal protein. Plant alkaloids, chlorophyll, and other porphyrins contain nitrogen in cyclic structures stable enough to have withstood changes during the coalification process and thus to have contributed to the nitrogen content of coal.

4.2.1 Test Method Protocols

The standard procedure of nitrogen determination by many laboratories is the Kjeldahl method (ASTM D-3179; ISO 333), although there are the standard methods that involve the Dumas technique (DIN 51722) and the gasification procedure (DIN 15722). Neutron activation analysis has also been proposed for the determination of nitrogen in coal, coal ash, and related products (Volborth, 1979a).

In the Kjeldahl method, pulverized coal is boiled coal with concentrated sulfuric acid containing potassium sulfate and a suitable catalyst to reduce the time for digestion. The catalyst is usually a mercury salt, selenium itself, or a selenium compound, or a mixture of the two. Selenium is regarded as being particularly advantageous.

The Kjeldahl–Gunning macro method is the one most widely used for determining nitrogen (ASTM D-3179). By this method, any nitrogen present in the sample is converted into ammonium salts by the destructive digestion of the sample by a hot mixture of concentrated sulfuric acid and potassium sulfate. After the digestion mixture has been made alkaline with sodium or potassium hydroxide, ammonia is expelled by distillation, condensed, and absorbed into a sulfuric acid solution and the excess acid is titrated with sodium hydroxide solution. The alternative method is similar except that after the complete digestion, the ammonia is distilled into a boric acid solution and titrated with a standard acid solution. Proper precautions should be taken in carrying out this procedure, especially the digesting and distillation steps. In addition to the possibility of losing nitrogen-containing species if the proper heating rate is not observed, there is the problem of working with hot concentrated sulfuric acid and caustic solutions.

A catalyst is used in the Kjeldahl–Gunning method to increase the rate of digestion of the coal and hence shorten the digestion period. For most bituminous and low-rank coals, the digestion period is on the order of 3 to 6 hours, even with the aid of a catalyst, and anthracite may require as much as 12 to 16 hours. A variety of catalysts can be used, including mercuric sulfate ($HgSO_4$) and selenium, mercuric selenite ($HgSeO_3$), or cupric selenite dihydrate ($CuSeO_4 \cdot 2H_2O$). However, when a mercury-containing catalyst is used, the addition of potassium or sodium sulfide (Na_2S) to the digestion mixture is necessary. The sulfide ions precipitate any mercuric ions as mercuric sulfide (HgS) and prevent formation of a stable mercury–ammonia complex ion (ammonia is produced in the reaction).

The most serious issue associated with the use of the Kjeldahl–Gunning method is whether or not complete conversion of nitrogen in the nitrogenous compounds to ammonia occurs. The main reason is that complete conversion may not be accomplished, for several reasons. The chemistry is more complex if not always straightforward, and in the presence of various oxidizing agents, nitrogen can be oxidized to nitrogen oxides which may be lost from the analysis. In addition, pyridine carboxylic acids may be formed, and these compounds are resistant to further decomposition. Thus, the somewhat limited oxidation and digestion conditions and the possible formation of unwanted but stable by-products usually require a lengthy digestion period. In many cases, to ensure complete conversion of the nitrogen to ammonia, the addition of chromic oxide (Cr_2O_3) to the digestion mixture increases the rate of digestion of coke.

The semimicro Kjeldahl–Gunning method (ASTM D-3179; ISO 333) has become a widely used method for determining nitrogen in coal, but there is some doubt about whether or not nitrogen recovery is complete by this procedure. In fact, the fate of nitrogen in the Kjeldahl method depends on its chemical form in the coal, the inorganic compounds added to catalyze the hydrolysis, and the amounts and types of compounds used to raise the boiling point of the mixture.

The Dumas method of nitrogen determination consists of an oxidation method in which a mixture of coal and copper oxide (CuO) is heated in an inert atmosphere to produce carbon dioxide (CO_2), water (H_2O), and nitrogen (N_2). The carbon dioxide is absorbed, the water condensed, and the nitrogen determined volumetrically. Although the Dumas method has been employed for many years, various modifications have been made to increase accuracy and precision.

The gasification method consists of mixing coal with a mixture of two parts of the Eschka mixture (i.e., 67% w/w light calcined magnesium oxide and 33% w/w anhydrous sodium carbonate), six parts of soda lime, and one part of molybdenum oxide. The sample is then placed in a porcelain boat, covered with platinum gauze, and heated (in a quartz tube) to 200 to 250°C (392 to 482°F). The sample is then heated in steam at 850 to 950°C (1562 to 1742°F). The gases pass into 0.1 *N* sulfuric acid, where ammonia is chemically absorbed as ammonium sulfate and then determined by one of the usual techniques.

Nitrogen determination by perchloric acid digestion of coal without catalysts is reportedly safe and much more rapid than the conventional Kjeldahl procedure. Nitrogen is determined in the digest by the usual ammonia evolution and titration. However, extreme care should be taken whenever perchloric acid is used. Nuclear reactions have been applied successfully to the determination of nitrogen in coal.

Nuclear reactions have been applied successfully to the determination of nitrogen in coal. Coal has been irradiated with fast neutrons from a 14-MeV generator to produce ^{13}N, which decays via positron emission and has a 10-minute half-life. Standard deviation for the nitrogen determinations was 0.07%. The instrumental neutron activation values were 6.8% higher (0.1% N) on average than those obtained using the standard Kjeldahl procedure. In the Kjeldahl procedure, loss of elemental nitrogen or nitrogen oxides or resistant heterocyclic nitrogen compounds can sometimes cause data errors, although most of these have been eliminated by application of modern laboratory procedures. Because results from instrumental neutron activation for nitrogen are independent of these factors, it has been suggested that the activation method is the more accurate of the two.

4.2.2 Data Handling and Interpretation

Nitrogen data are used primarily in research and for the comparison of coals. These values are needed so that the oxygen content of a coal can be estimated by difference. During combustion, the nitrogen in coal can be converted to ammonia, elemental nitrogen, or nitrogen oxides, depending on the conditions of burning and the nature of the coal used. Nitrogen values could possibly be used to estimate the amount of nitrogen oxides that would be emitted upon burning of certain coals. Coal nitrogen values are also useful in predicting the amount of nitrogen in the products of coal liquefaction and gasification processes.

4.3 SULFUR

Sulfur is an important consideration in coal utilization, and hence, there is a considerable amount of published work relating to the development of methods to improve the efficiency of the techniques as well as improve the accuracy and precision of the sulfur determination (Ahmed and Whalley, 1978; Chakrabarti, 1978a; Attar, 1979; Raymond, 1982; Gorbaty et al., 1992).

Total sulfur data (ASTM D-3177; ASTM D-4239) are necessary for the effective control of the emissions of oxides of sulfur whenever coal is used as a fuel.

The emission of sulfur oxides leads to the corrosion of equipment and slagging of combustion or boiler equipment, as well as contributing to atmospheric pollution and environmental damage. Sulfur data are therefore necessary for the evaluation of coals to be used for combustion purposes.

Most coal conversion and cleaning processes require two sets of sulfur values: the sulfur content of the coal before it is used and the sulfur content of the products formed. In the coking of coal, some of the sulfur is removed in the coking process, which makes it necessary to obtain the before and after values. The commercial uses of coke, as in metallurgical processes, require a low sulfur content and necessitate an accurate sulfur value for the coke. In coal gasification and liquefaction processes, the sulfur in the coal is sometimes carried through to the products. It is therefore necessary to determine the amount of sulfur in each of the products before it is used. One of the primary reasons for cleaning coal is to reduce the sulfur content. It is necessary to know the sulfur content before and after cleaning in order to evaluate the cleaning process.

Total sulfur values alone are not adequate in accessing a cleaning process for reducing the sulfur content of coal. Pyrite sulfur alone can be removed by specific gravity separations, and its removal depends on the way the pyrite is distributed throughout the coal. If it occurs as very small crystals widely dispersed in the coal, it is almost impossible to remove by these methods. When pyrite occurs in large pieces, it can be removed successfully by specific gravity methods. Organic sulfur cannot be reduced appreciably, since it is usually uniformly dispersed throughout the organic material in coal.

Sulfur is present in coal in three forms, either as (1) organically bound sulfur, (2) inorganic sulfur (pyrite or marcasite, FeS_2), or (3) inorganic sulfates (ASTM D-2492; ISO 157; Kuhn, 1977). The amount of organic sulfur is usually <3% w/w of the coal, although exceptionally high amounts of sulfur (up to 11%) have been recorded. Sulfates (mainly calcium sulfate, $CaSO_4$, and iron sulfate, $FeSO_4$) rarely exceed 0.1% except in highly weathered or oxidized samples of coal. Pyrite and marcasite (the two common crystal forms of FeS_2) are difficult to distinguish from one another and are often (incorrectly) designated simply as pyrite. Free sulfur as such does not occur in coal to any significant extent. The amount of the sulfur-containing materials in coal varies considerably, especially for coals from different seams. In addition, pyrite is not uniformly distributed in coal and can occur as layers or slabs or may be disseminated throughout the organic material as very fine crystals. The content of sulfates, mainly gypsum ($CaSO_4 \cdot 7H_2O$) and ferrous sulfate ($FeSO_4 \cdot 7H_2O$), rarely exceeds a few hundredths of a percent, except in highly weathered or oxidized coals.

4.3.1 Test Method Protocols

The three most widely used test methods for sulfur determination are (1) the Eschka method, (2) the bomb washing method, and (3) the high-temperature combustion method, and all are based on the combustion of the sulfur-containing material to produce sulfate, which can be measured either gravimetrically or volumetrically. The Eschka method has distinct advantages in that the equipment is relatively simple and only the more convenient analytical techniques are employed. However, methods involving combustion of the organic material in a *bomb* have distinct advantages insofar as sulfur is not *lost* during the process, and such methods are particularly favored when the calorific value of the coal is also required. The method of decomposition involves the Parr fusion procedure in the presence of sodium peroxide and oxygen at high pressures [300 to 450 psi (2.1 to 3.1 MPa)].

In the *Eschka method* (ASTM D-3177; ISO 334; ISO 351), 1 g of the analysis sample is thoroughly mixed with 3 g of Eschka mixture, which is a combination of two parts by weight of light calcined magnesium oxide with one part of anhydrous sodium carbonate. The combination of sample and Eschka mixture is placed in a porcelain crucible (30 mL) and covered with another gram of Eschka mixture. The crucible is placed in a muffle furnace, heated to a temperature of $800 \pm 25°C$ ($1472 \pm 45°F$), and held at this temperature until oxidation of the sample is complete. The sulfur compounds evolved during combustion react with the magnesium oxide (MgO) and sodium carbonate (Na_2CO_3) and under oxidizing conditions are retained as magnesium sulfate ($MgSO_4$) and sodium sulfate (Na_2SO_4). The sulfate in the residue is extracted and determined gravimetrically.

In the *bomb washing method*, sulfur is determined in the washings from the oxygen bomb calorimeter following the calorimetric determination (ASTM D-2015 and ASTM D-3286; these standards have been discontinued but are still used in many laboratories). After opening, the inside of the bomb is washed carefully, and the washings are collected. After titration with standard base solution to determine the acid correction for the heating value, the solution is heated and treated with ammonium hydroxide (NH_4OH) to precipitate iron ions as ferric oxide [$Fe(OH)_3$]. After filtering and heating, the sulfate is precipitated with barium chloride ($BaCl_2$) and determined gravimetrically.

In the *high-temperature combustion test method* (ASTM D-4239, submethods A, B, and C), a weighed sample is burned in a tube furnace at a minimum operating temperature of 1350°C (2462°F) in a stream of oxygen to ensure complete oxidation of sulfur-containing components in the sample. Using these conditions, all sulfur-containing materials in the coal or coke are converted predominantly to sulfur dioxide in a reproducible way. The amount of sulfur dioxide produced by burning the sample can be determined by the three alternative methods mentioned above.

In the acid–base titration method, the combustion gases are bubbled through a hydrogen peroxide solution in a gas absorption bulb. Sulfuric acid is produced by the reaction of sulfur dioxide with hydrogen peroxide and is determined by titration with a standard base solution. Chlorine-containing species in the sample yield hydrochloric acid in the hydrogen peroxide solution, which contributes to its total acidity. For accurate results, a correction must be made for the chlorine present in the sample (ASTM D-2361; ASTM D-4208; ISO 352; ISO 587). Appropriate standard reference materials should be used to calibrate commercially available sulfur analyzers to establish recovery factors or a calibration curve based on the range of sulfur in the coal or coke samples being analyzed.

In the iodimetric titration procedure, the combustion gases are bubbled through a diluent solution containing pyridine, methanol, and water. This solution is titrated with a titrant containing iodine in a pyridine, methanol, and water solution. In automated systems, the titrant is delivered automatically from a calibrated burette syringe and the endpoint detected amperometrically. The method is empirical, and standard reference materials with sulfur percentages in the range of the samples to be analyzed should be used to calibrate the instrument before use. Alternative formulations for the diluent and titrant may be used in this method to the extent that they can be demonstrated to yield equivalent results.

The third method of measuring the sulfur dioxide in the combustion gases is by the absorption of infrared (IR) radiation. Moisture and particulates are first removed from the gas stream by traps filled with anhydrous magnesium perchlorate. The gas stream is then passed through an infrared absorption cell tuned to a frequency of radiation absorbed by sulfur dioxide. The infrared radiation absorbed during combustion of the sample is proportional to the sulfur dioxide in the combustion gases and therefore to the sulfur in the sample. The method is empirical, and standard reference materials with sulfur percentages in the range of the samples to be analyzed should be used to calibrate the instrument before use.

Some general problems associated with the determination of sulfur in coal are nonuniform distribution of pyrite particles, failure to convert all the sulfur to sulfate, and loss of sulfur as sulfur dioxide during the analysis. The nonuniform distribution of pyrite necessitates the collection of many sample increments to ensure that the gross sample is representative of the lot of coal in question. Pyrite particles are both hard and heavy and have a tendency to segregate during the preparation and handling of samples. Because the particles are harder, they are more difficult to crush and pulverize and tend to concentrate in the last portion of material that remains from these processes.

In each of the methods discussed, sulfur is oxidized to sulfur dioxide during the analysis. Some sulfur dioxide may be lost unless the necessary precautions are taken. In the Eschka method, a generous layer of Eschka mixture covering the fusion mixture helps prevent the loss of sulfur as sulfur dioxide. The mixture must be heated gradually to guard against the production of sulfur dioxide at a rate that is too high for it to be absorbed by the Eschka mixture. In the bomb washing method, the pressure of the bomb should be released slowly after the sample is burned in oxygen so that sulfur oxides will not be carried out of the bomb. In the high-temperature combustion methods, it is essential that the flow of oxygen be sufficient and that the rate of heating not be too high. A high rate of heating will lead to the evolution of combustion products, including sulfur dioxide, at a rate that is too rapid for complete absorption in the solutions or for detection by the infrared cell.

The gravimetric determination of sulfate can be and is most often used to finish the Eschka and bomb washing methods. The most serious concern is that the barium sulfate precipitated may be extremely fine and difficult to filter. This can be overcome by adding the barium chloride ($BaCl_2$) rapidly to the hot solution and stirring the mixture vigorously to obtain a barium sulfate ($BaSO_4$) precipitate, which is easily filtered out. In addition, heating and digestion for a lengthy period improve the filterability of the precipitate. After filtering, the precipitate must be washed several times with hot water to remove adsorbed materials that will cause the results to be too high.

4.3.2 Determination of the Forms of Sulfur

As noted above, sulfur occurs in coal in three forms (ASTM D-2492; ISO 157): (1) inorganic sulfur or sulfate sulfur [i.e., as sulfates of metals such as calcium sulfate ($CaSO_4$)], (2) pyrite sulfur [i.e., as pyrite or marcasite (FeS_2)], and (3) organic sulfur (i.e., as sulfur combined organically in the coal). Organic sulfur and pyrites account for almost all the sulfur in coal. Sulfate sulfur is usually less than 0.1%, except in weathered coal containing an appreciable amount of pyrites. The pyrite sulfur content varies considerably more than does the organic sulfur content and is of more interest because it is the form that can most easily be removed from coal by current preparation practices.

The methods used for the determination of the various forms of sulfur in coal are based on the fact that sulfate sulfur is soluble in dilute hydrochloric acid solution, whereas pyrite and organic sulfur are not attacked. In addition, dilute nitric acid dissolves sulfate and pyrite sulfur quantitatively and attacks organic sulfur only slightly. Pyrite sulfur, however, is determined accurately by noting the quantity of iron extracted by dilute hydrochloric acid and subtracting this from the iron extracted by nitric acid, the difference being the iron present as pyrite iron (FeS_2), from which the equivalent quantity of pyrite sulfur can be calculated. Sulfate sulfur and pyrite sulfur may therefore be determined on the one sample of coal, but since the sulfate sulfur determination is rather tedious, it is customary to use separate samples for each determination. Organic sulfur is then obtained by subtracting the combined percentages of sulfate sulfur and pyrite sulfur from the total sulfur determined by the Eschka method.

In the method for determining the forms of sulfur in coal (ASTM D-2492; ISO 157), the sulfate and pyrite sulfur are determined directly, and the organic sulfur is taken as the difference between the total sulfur and the sum of the sulfate and pyrite sulfur.

In the determination of sulfate, 2 to 5 g of the analysis sample is mixed with HCl (2 volumes concentrated HCl + 3 volumes of water), and the mixture is gently boiled for 30 minutes. After filtering and washing, the undissolved coal may be retained for the determination of pyrite sulfur, or it may be discarded and a fresh sample used for pyrite sulfur. Saturated bromine water is added to the filtrate to oxidize all sulfur forms to sulfate ions and ferrous ions to ferric ions. After boiling to remove excess bromine, the iron is precipitated with excess ammonia and filtered. This precipitate must be retained for the determination of nonpyrite iron if a fresh sample of coal was used for the determination of the pyrite iron. The sulfate is then precipitated with $BaCl_2$, and the $BaSO_4$ is determined gravimetrically.

Either the residue from the sulfate determination or a fresh 1-g sample is used for the determination of pyrite sulfur content. The sample is added to dilute nitric acid and the mixture boiled gently for 30 minutes or allowed to stand overnight. This treatment oxidizes iron species to iron(III) and inorganic sulfur compounds to sulfate. The mixture is then filtered, and the filtrate is saved for the determination of iron by atomic absorption spectrophotometry or by a titration procedure. If iron is to be determined by the atomic absorption method, no further work is done on the filtrate other than to dilute it to an appropriate volume before the determination. If a titration method is to be used for the determination of iron, the filtrate is treated with 30% H_2O_2 to destroy any coloration arising from the coal. The iron is then precipitated, filtered, and washed. The precipitate is then dissolved in HCl, and the iron determined by titration with either potassium dichromate ($K_2Cr_2O_7$) or potassium permanganate ($KMnO_4$).

If a new sample is used for the determination of pyrite iron, the iron determined by these procedures represents the combination of the pyrite and nonpyrite iron. The amount of nonpyrite iron must then be determined separately and subtracted from the amount determined. If the residue from the sulfate determination was used, the iron determined by the foregoing procedures represents the pyrite iron. Once the correct value for the pyrite iron is determined, the pyrite sulfur is calculated using the following expression:

$$\%\text{ pyrite sulfur} = \%\text{ pyrite iron} \times 2 \times \frac{32.06}{55.85} \tag{4.2}$$

where $2 \times 32.06/55.85$ is the ratio of sulfur to iron in pyrite.

Some difficulties encountered in determining the amounts of the various forms of sulfur in coal are adsorption of other materials on barium sulfate ($BaSO_4$) when it is precipitated, inability to extract all the pyrite sulfur from the coal during the extraction process, and possible oxidation of pyrite sulfur to sulfate in the pulverization and storage of the coal sample. Both the adsorption of other materials on barium sulfate and oxidation of pyrite sulfur lead to high values for the sulfate sulfur. Iron ions (Fe^{2+}) are readily adsorbed on barium sulfate, which could be particularly objectionable for coals containing large amounts of nonpyrite iron. Removal of the iron by precipitation and filtration before the precipitation of barium sulfate minimizes adsorption of the iron. Inadequate pulverization and mixing of the sample appear to be the major causes of the incomplete extraction of pyrite sulfur from coal. A very small amount of organic sulfur may also be extracted with the pyrite sulfur. For this reason, the amount of pyrite iron extracted is used as a measure of the pyrite sulfur. To control the oxidation of pyrite sulfur to sulfates, exposure of the coal sample to the atmosphere at elevated temperatures should be avoided, and the sample should be analyzed as soon as possible.

X-ray fluorescence has been used extensively for determining total sulfur in coal. The economy and speed of such x-ray fluorescence methods when used for multiple determinations (e.g., Al, Si, Ca, Mg, Fie, K, Ti, P, and S) in the same prepared coal sample are probably unsurpassed by any other method.

A neutron sulfur meter was developed for continuous monitoring of the sulfur content of coal from a preparation plant. Neutrons from a radioactive source produced thermal neutrons within the sample, which produced prompt gamma-ray emissions when captured by atoms in the coal. The gamma-ray energies characteristic of sulfur were measured by a sodium iodide crystal and associated electronic equipment. Several methods to eliminate interferences from other elements, moisture, and bulk density were evaluated. Results from the neutron sulfur meter for total sulfur in coal agreed closely with chemical analyses, and the method is precise (±0.05% absolute), rapid, and suitable for online sulfur determination.

High-resolution x-ray spectroscopy was employed successfully to determine the chemical state of sulfur in coal. Fluorescent x-rays from pressed coal powders, excited with x-rays from a chromium target tube, were analyzed using a germanium crystal. The sulfur K_a radiation increased with increasing oxidation number, allowing quantitative determinations of the sulfur types.

Another method of direct determination of organic sulfur is to subject a small (20- to 30-mg) sample of <200-mesh coal to low-temperature ashing 1 to 3 hours and to collect the sulfur oxides evolved in a cold trap. They are then absorbed in hydrogen peroxide (H_2O_2) and determined chromatographically.

Mossbauer spectrometry is also useful to identify multiple iron species in coal and charred residues without using a concentration step. The results indicate that heat treatment of any kind, even at temperatures as low as 175°C (347°F), changes the nature of the iron species in coal. Furthermore, some kind of an association between the pyrite (FeS_2) in whole coal and the organic matrix was indicated. The pyrite also appeared to be altered when the coal was heated to 175°C (347°F).

4.3.3 Data Handling and Interpretation

The principal use of forms of sulfur data is in connection with the cleaning of coal. Within certain limits, pyrite sulfur can be removed from coal by gravity separation methods, whereas organic sulfur cannot. Pyrite sulfur content can therefore be used to predict how much sulfur can be removed from the coal and to evaluate cleaning processes. If the pyrite sulfur occurs in layers, it can usually be removed efficiently. If it occurs as fine crystals dispersed throughout the coal, its removal is very difficult.

Other uses of forms of sulfur data are the inclusion of the pyrite sulfur value in the formula for the estimation of oxygen by difference and as a possible means of predicting the extent of weathering of coal. The sulfate concentration increases upon weathering, so the sulfate sulfur value could be used as an indication of the extent of weathering of coal.

4.4 OXYGEN

Oxygen occurs in both the organic and inorganic portions of coal. In the organic portion, oxygen is present in hydroxyl (–OH), usually phenol groups, carboxyl groups (CO_2H), methoxyl groups ($-OCH_3$), and carbonyl groups ($=C=O$). In low-rank coal, the hydroxyl oxygen averages about 6 to 9%, whereas high-rank coals contain less than 1%. The percentages of oxygen in carbonyl, methoxyl, and carboxyl groups average from a few percent in low rank and brown coal to almost no measurable value in high-rank coal. The inorganic materials in coal that contain oxygen are the various forms of moisture, silicates, carbonates, oxides, and sulfates. The silicates are primarily aluminum silicates found in the shalelike portions. Most of the carbonate is calcium carbonate ($CaCO_3$), the oxides are mainly iron oxides (FeO and Fe_2O_3), and the sulfates are calcium and iron ($CaSO_4$ and $FeSO_4$).

4.4.1 Test Method Protocols

For many years, no satisfactory method existed for the direct determination of oxygen in coal. In practice, in the expression of the ultimate analysis it is customary to deduct from 100.0 the sum of the percentages of moisture, ash, carbon, hydrogen, nitrogen, and sulfur (ASTM D-3176):

$$\%\ \text{oxygen} = 100 - (\%\text{C} + \%\text{H} + \%\text{N} + \%\text{S}_{\text{organic}}) \tag{4.3}$$

The resulting figure for oxygen is therefore burdened with the accumulation of all the experimental errors involved in the determinations of the other constituents that form part of the equation. Consequently, the oxygen content was, and still is, of a low order of accuracy. For this reason, oxygen determined by difference should not be designated *percent oxygen* but *percent oxygen by difference*.

The most widely used procedure for oxygen determination consists of pyrolyzing coal in the presence of nitrogen and subsequent passage of the products over hot (1100°C, 2012°F) carbon or platinized carbon (ISO, 1994). The oxygen in the volatile products is thereby converted to carbon monoxide, which can be determined by a variety of techniques (Gluskoter et al., 1981; Frigge, 1984). For example, the carbon monoxide can be oxidized to carbon dioxide, usually with iodine pentoxide, which releases free iodine. The iodine released can be determined titrimetrically, or the carbon dioxide produced can be absorbed and determined gravimetrically to calculate the amount of oxygen in the original samples.

However, the various sources of oxygen in coal, such as the oxygen in the moisture and water of hydration of mineral matter, the oxygen in carbonates, and the oxygen in silicates and other inorganic compounds, in addition to the oxygen in the organic matter, all offer hurdles in producing data that are beyond question. The original procedure has been modified in several ways to reduce the contribution made by some of these oxygen sources to the determined oxygen value. Thorough drying in a nitrogen atmosphere before pyrolysis of the sample minimizes the effect of moisture, and much of the mineral matter is removed by a specific gravity separation or chemical treatment with hydrochloric and hydrofluoric acid. The reduction of mineral matter minimizes the contribution that the water of hydration and the inorganic compounds, such as carbonates, silicates, oxides, and sulfates, make to the oxygen value that is determined. After all the pretreatment steps are taken to remove moisture and mineral matter, the oxygen value obtained by this method is essentially a measure of the oxygen contained in the organic matter in coal.

Several other methods for the direct determination of oxygen have met with some success when applied to coal and therefore deserve mention here, because it is conceivable that at some future date one of these methods (or a modification thereof) could find approval as a recognized standard method for the direct determination of oxygen in coal.

An oxidation method that has been applied to coal involves the combustion of a weighed amount of coal with a specific quantity of oxygen. The oxygen in the coal is determined by deducting the added quantity of oxygen from the sum of the residual oxygen and the oxygen of the oxidation products. The disadvantages of the method arise because of the difficulties associated with the accurate measurement of the substantial volume of oxygen used and the quantity and quality (i.e., composition) of the combustion products.

Another method for the determination of oxygen in coal involves reduction of the coal by pyrolysis in the presence of hydrogen, whereupon the oxygen is converted catalytically to water. However, the procedure is relatively complex and the catalyst may be poisoned by sulfur and by chlorine.

In the neutron activation method (Volborth, 1979a,b; Volborth et al., 1987; Mahajan, 1985), as illustrated by the reaction

$$16\text{O} + \text{neutron} = 16\text{N} + \text{proton} \qquad (4.4)$$

the concentration of oxygen is determined by measuring the radiation from the sample. The method is nondestructive and rapid, but if only the organic oxygen is to be determined, the sample must first be demineralized.

Because oxygen in coal includes both oxygen in organic matter and inorganic oxygen from mineral matter and moisture, demineralization or float–sink cleaning of coal is usually employed to remove inorganic oxygen. Using a modified method (ISO 602; ISO, 1994), the results of the direct determination of oxygen in untreated coals showed significant variations from oxygen values obtained by the difference method. The differences were caused primarily by sulfur trioxide being retained in the ash, sulfur in the pyrite form in the whole coal, and small quantities of inorganic oxygen. Much closer agreement was obtained when ash values used for calculation of oxygen by difference were corrected to a better approximation of mineral matter:

$$\text{corrected ash} = \text{ash}_{\text{determlned}} - \text{SO}_{3\text{ash}} \qquad (4.5)$$

Pyrite sulfur in whole coal is added to the ash value, organic sulfur is added in the ultimate analysis as part of the coal substance, and CO from carbonates is added to total carbon and oxygen values. Demineralized coals gave the best results for organic oxygen; values from untreated coals were usually higher because they included inorganic oxygen.

The use of gas chromatography to determine oxygen in coal is considerably faster than the methods just described. The oxygen produced from coal pyrolysis in a vacuum was converted to carbon monoxide catalytically, and the total quantity of gas was measured. Gas chromatography was then used to measure the concentration of carbon monoxide in the gas.

A method employing reduction fusion using radio-frequency heating of coal has been reported for the determination of oxygen in coal. An induction furnace was used to heat an iron–tin bath in a graphite crucible supported in a bed of carbon black powder that was contained in a silica thimble. All were enclosed in a reaction tube through which argon carrier gas flowed. Pyrolysis products from a sealed coal sample (10 mg) dropped into the bath were purified, and oxygen was converted to carbon dioxide that was absorbed and weighed. Satisfactory results were obtained for organic oxygen in demineralized coal. For untreated coal, satisfactory results were obtained when a range of correction techniques for oxygen was applied to coals containing more than 5% ash.

4.4.2 Data Handling and Interpretation

When the oxygen value is estimated by subtracting from 100 the percentages determined for all other constituents, the errors in the values determined are reflected in the oxygen value estimated. These errors may be partially compensating, or they may be additive. It is important that accurate determinations be made and appropriate corrections for overlapping values, especially hydrogen, be calculated. Thus,

$$\mathrm{O} = 100 - (\mathrm{C} + \mathrm{H} + \mathrm{N} + \mathrm{S} + \mathrm{ash}) \tag{4.6}$$

All values are expressed as percentages and C is the as-determined carbon (ASTM D-3178), H is the as-determined hydrogen (ASTM D-3178), N is the as-determined nitrogen (ASTM D-3179), S is the as-determined sulfur (ASTM D-3177; ASTM D-4239), and "ash" is the as-determined ash (ASTM D-3174). All the values above pertain to those obtained from the analysis sample. The as-determined value for carbon represents both organic and carbonate carbon. The as-determined value for hydrogen represents the organic hydrogen, the hydrogen in the residual moisture, and the hydrogen in the water of hydration of mineral matter. The sulfur value is the total sulfur in coal and represents that which is contained in the organic matter, pyrites, and sulfates. Ash is mostly metal oxides and silicon oxide. The estimated value of the oxygen therefore includes the oxygen contained in the organic matter, in the moisture, and in the mineral matter, except that which is combined with metals in coal ash.

A rough estimate of the oxygen contained in the organic matter in coal can be obtained by correcting the oxygen value obtained by equation (4.6) for the oxygen in residual moisture and water of hydration of mineral matter. Adding this correction to the sum subtracted from 100 gives the following expression:

$$\mathrm{O}_x = 100 - [\mathrm{C} + \mathrm{H} + \mathrm{N} + \mathrm{S} + \mathrm{ash} + \tfrac{8}{9}(\mathrm{H_2O} + \mathrm{H_2O\ of\ hydration})] \tag{4.7}$$

where H_2O is the as-determined residual moisture (ASTM D-3173) and the H_2O of hydration is 8.0% of ash.

Several improvements in the estimation of organic oxygen can be made when analytical data are available. Values for chlorine, carbon dioxide, pyrite sulfur, and sulfur in coal ash are helpful in improving the estimation. Failure to include the chlorine value in the sum subtracted from 100 leads to a high value for the organic oxygen. This oxygen value should also be reduced for the oxygen present in the carbon dioxide that is associated with the mineral matter.

The inclusion of both total sulfur and ash in the sum that is subtracted from 100% in estimating the oxygen content of coal lowers the oxygen value, since part of the sulfur may be retained in the ash. The sulfur that is retained is therefore counted twice in the sum for subtraction. Correcting the coal ash for the sulfur trioxide (SO_3) present compensates for this error. Similarly, the coal ash should be corrected for any ferric oxide (Fe_2O_3) that results from the heating of pyrite (FeS_2) in air, as is done in the ashing process. In the ashing process, three oxygen atoms replace four sulfur atoms:

$$4FeS_2 + 11O_2 \rightarrow 2Fe_2O_3 + 8SO_2 \tag{4.8}$$

On a weight basis, 48 parts of oxygen replace 128 parts of sulfur. This oxygen is from an external source and not from the coal itself. Since this oxygen contributes to the weight of the ash, correction for the pyrite sulfur value is necessary. The pyrite sulfur that is replaced is accounted for in the total sulfur value. When the values are available to make these corrections, a good estimate can be made of the percentage of oxygen in the organic or combustible portion of coal:

$$\begin{aligned} O = 100 - [&C + H + N + S + Cl + (\text{ash} - \%S_p - SO_{3\text{ash}}) \\ &+ \tfrac{8}{9}(H_2O + H_2O_{\text{hydration}}) + \tfrac{32}{44}CO_2] \end{aligned} \tag{4.9}$$

Chlorine is the as-determined chlorine (ASTM D-2361; ASTM D-4208), S_p is the as-determined pyrite sulfur (ASTM D-2492), $SO_{3\text{ash}}$ is the as-determined sulfate (sulfur trioxide, SO_3) in ash (ASTM D-1757), and CO_2 is the as-determined carbon dioxide in coal (ASTM D-1756). All other terms are as given in the earlier formulas, and all values are expressed as percentages.

If the oxygen is determined directly, using the method discussed previously in which the moisture and mineral matter are removed from the sample before actual oxygen determination, the oxygen value in this case represents that contained in the organic matter. In calculating heat balances for boiler efficiency studies, it is important that an accurate value of the combustible material in coal be obtained. Thus, a correction for the oxygen content of the organic matter of coal should be made. Of course, corrections to the carbon and hydrogen values for the amount of these elements found in the moisture and inorganic constituents of coal should also be made. Oxygen data are used for determining the suitability of coals for coking, liquefaction, or gasification processes. In general, coals with a high oxygen content are unsuitable for coking but may be more reactive and thus easier to gasify or liquefy.

4.5 CHLORINE

Chlorine occurs in coal (Chakrabarti, 1978, 1982; Hower et al., 1992) and is believed to be a factor not only in fouling problems but also in corrosion problems (Canfield et al., 1979; Slack, 1981). The chlorine content of coal is normally low, usually only a few tenths of a percent or less. It occurs predominantly as sodium, potassium, and calcium chlorides, with magnesium and iron chlorides present in some coals. There is evidence that chlorine may also be combined with the organic matter in coal.

4.5.1 Test Method Protocols

Methods of converting the chlorine in coal into a form suitable for its analytical determination include combusting the sample, with or without Eschka mixture, in an oxygen bomb and heating with Eschka mixture in an oxidizing atmosphere. Eschka mixture is a combination of two parts by weight of magnesium oxide and one part of anhydrous sodium carbonate. There are two standard methods of determining chlorine in coal (ASTM D-2361; ASTM D-4208).

The first method (ASTM D-2361) offers a choice of two procedures for combusting the coal sample. In the bomb combustion procedure, the oxygen bomb used is the same as, or very similar to, that used in determination of the calorific value (ASTM D-2015; ASTM D-3286).

In the first procedure (ASTM D-2361), a weighed sample (approximately 1 g) is mixed with a weight amount (approximately 1 g) of the Eschka mixture and placed in a crucible inside an oxygen bomb. An ammonium carbonate solution is added to the bomb to trap the chloride-containing species produced in the combustion. After charging with oxygen to 25 atm, the bomb is fired and allowed to stand in the calorimeter water for at least 10 minutes. The pressure on the bomb is then released slowly, the bomb is disassembled, and all parts of the bomb interior are washed with hot water. The washings are collected in a beaker and acidified with nitric acid. The amount of chloride in the solution is then determined by either a modified Volhard or by a potentiometric titration with silver nitrate solution. In the second procedure (ASTM D-2361), 1 g of the coal analysis sample is mixed with 3 g of Eschka mixture in a suitable crucible. This mixture is covered with an additional 2 g of Eschka mixture to ensure that no chlorine is lost during combustion. The mixture is then ignited gradually in a muffle furnace by raising the temperature to 675 ± 25°C within 1 hour. This temperature is maintained for 1 hour before the cooling and washing of the incinerated mixture with hot water into a beaker. The contents of the beaker are acidified with nitric acid, and the chloride is determined as in the procedure described previously.

In second test method (ASTM D-4208), 1 g of the analysis sample of coal is placed in a crucible inside an oxygen bomb. A sodium carbonate solution is added to the bomb to trap the chloride-containing species produced. After charging with oxygen to 25 atm, the bomb is fired and allowed to stand in the calorimeter water for at least 15 minutes. After the pressure is released very slowly, the bomb is disassembled, and all parts of the bomb interior are washed with water. The washings are collected, an ionic strength adjuster (sodium nitrate, $NaNO_3$) is added, and the chloride is determined with an ion-selective electrode by the standard addition method.

In both methods it is possible to lose some of the chlorine during combustion unless necessary precautions are taken. Thoroughly mixing the coal sample with Eschka mixture and covering this carefully with additional Eschka mixture minimizes the loss of chlorine. In bomb combustion methods, the ammonium and sodium carbonate solutions in the bomb are used to absorb the chlorine as it is released in the combustion. The 10- and 15-minute waiting periods and the slow release of pressure on the bomb help to prevent the loss of chlorine as well.

A modification of the oxygen bomb combustion method (ASTM D-2361) for the determination of chlorine consisted of acidifying a solution of the adsorbed combustion products and titrating the chlorine potentiometrically. A potentiometric titration was also tried for the determination of chlorine by the oxygen flask method. Combustion products, including chlorine, were absorbed in sodium hydroxide (NaOH), and the chloride was measured using silver–silver chloride electrodes. Although there was no statistical difference in results obtained from potentiometric titration and the Eschka procedure, the latter was more precise.

There is also a test method (ASTM E-256) for the determination of chlorine in organic compounds by sodium peroxide bomb ignition that is also worthy of reference. The method is intended for application to samples of organic materials containing more than 0.5% chlorine, and the assumption is that halogens other than chlorine will not be present.

4.5.2 Data Handling and Interpretation

The chlorine in coal and in the products derived from coal is known to contribute significantly to corrosion of the coal-handling and coal-processing equipment. Since the corrosion of this equipment is the result of several causes, one being the chlorine content of coal, it is difficult to predict the degree of corrosion within a given time frame. It is equally difficult to predict the degree to which the chlorine content contributes to the corrosion, other than the general prediction that the higher the chlorine content, the greater the chances for corrosion of the equipment. As a general rule, coal with a high chlorine content is not desirable.

Chlorine data are used in ultimate analysis to improve the estimate of oxygen by difference. The chlorine value is included in the sum of the items determined, which, when subtracted from 100, gives an estimate of the oxygen content of coal.

Although the validity of the analytical data is uncertain, a generally accepted fouling classification of coal, according to total chlorine content (ASTM D-2361; ISO 352; ISO 587), is as follows:

Total Percent Chlorine in Coal	Fouling Type
<0.2	Low
0.2–0.3	Medium
0.3–0.5	High
>0.5	Severe

In terms of corrosion, the occurrence of chlorine in coal leads to the formation of hydrogen chloride, and the condensation of water containing hydrogen chloride (hydrochloric acid) on the cooler parts of combustion equipment can lead to severe corrosion of the metal surfaces.

4.6 MERCURY

Mercury has been identified as a very dangerous environmental contaminant, largely by reason of the process of concentration in the food chain. Thus, the presence of mercury in coal is an extremely sensitive issue. The possible emission of mercury that may be found in coal is an environmental concern.

The test for total mercury (ASTM D-3684; ISO 15237) involves combusting a weighed sample in an oxygen bomb with dilute nitric acid absorbing the mercury vapors. The bomb is rinsed into a reduction vessel with dilute nitric acid, and the mercury is determined by the flameless cold vapor atomic absorption technique. Mercury and mercury salts can be volatilized at low temperatures. Precautions against inadvertent mercury loss should be taken when using this method. When coal samples are burned according to this procedure, provided that sample preparation in is accord with the standard procedure (ASTM D-2013), the total mercury is quantitatively retained and is representative of concentrations in the whole coal. Caution is also advised in use of the bomb, considering the chemicals required for the test method.

Another test method for the determination of mercury in coal (ASTM D-6414) involves (method A) solubilizing of the mercury in the sample by heating the sample at a specified temperature in a mixture of nitric and hydrochloric acids. The acid solutions produced are transferred into a vessel in which the mercury is reduced to elemental mercury. The mercury vapor is determined by flameless cold-vapor atomic absorption spectroscopy. An alternative method (method B) involved solubilization of the mercury by heating the sample in a mixture of nitric acid and sulfuric acid with vanadium pentoxide. The acid solution is then transferred into a vessel in which the mercury is reduced to elemental mercury. The mercury content is determined by flameless cold-vapor atomic absorption spectroscopy. However, mercury and mercury salts can be volatilized at low temperatures, and precautions against inadvertent mercury loss should be taken when using this method.

The determination of mercury in coal, and in coal combustion residues, can also be accomplished by controlled heating of the sample in oxygen (ASTM

D-6722). The sample is heated to dryness and then decomposed thermally and chemically. The decomposition products are carried by flowing oxygen to the catalytic section of the furnace, where oxidation is completed and halogens as well as nitrogen and sulfur oxides are trapped. The remaining decomposition products are carried to a gold amalgamator that traps mercury selectively. After the system is flushed with oxygen to remove any remaining decomposition products, the amalgamator is heated rapidly, releasing mercury vapor. Flowing oxygen carries the mercury vapor through absorbance cells positioned in the light path of a single-wavelength atomic absorption spectrophotometer. Absorbance peak height or peak area, as a function of mercury concentration, is measured at 253.7 nm. As before, precautions must be taken against loss of mercury and mercury salts that are volatile at low temperatures.

4.7 OTHER CONSTITUENTS

Several of the minor components of coal are of importance, because of the quantity present on occasion, but more so in some cases by virtue of the special properties they possess which are undesirable when the coal is used for certain purposes. For example, to arrive at a correct figure for the combustible carbon in coal, it is necessary to apply a correction for the quantity of carbonate associated with the sample. Combustion analyses determine only the total carbon. Again, coking coals should have low phosphorus content, and anthracites used for malting should contain only very small quantities of arsenic, so that the determination of these elements becomes necessary in certain cases. Since both are found normally in small amounts, they are not included in the general statement of the ultimate analysis but are reported separately.

4.7.1 Carbon Dioxide

Most coals contain small amounts of mineral carbonates made up primarily of calcium carbonate and to a lesser extent ferrous and other metal carbonates. Some coals contain a comparatively large amount of the inorganic carbonates, and the determination of carbon dioxide content is required in estimating the mineral matter content of these high-carbonate coals. Indeed, it is necessary to have a knowledge of the carbonate content of coal to correct the carbon figure and since, without resorting to very elaborate analyses, it would be impossible to express the carbonate content as definite quantities of calcium carbonate, magnesium carbonate, and so on, it is customary, and sufficient for all analytical purposes, to express it simply in terms of carbon dioxide.

One particular test method (ASTM D-1756) covers the determination of carbon dioxide in coal in any form, such as mineral carbonate, from which carbon dioxide is released by action of mineral acids (e.g., hydrochloric acid). The method can be applied to high- and low-carbonate coals. The determination of carbon dioxide is made by decomposing with acid a weighed quantity of the sample in a closed system and absorbing the carbon dioxide in an absorbent (e.g., such as sodium hydroxide, NaOH, or potassium hydroxide, KOH). The increase in weight of the absorbent is a measure of the carbon dioxide in the sample, which can be used to calculate the amount of mineral carbonates in the coal.

Due to the small amount of carbon dioxide in coal and the difficulty of measuring accurately the carbon dioxide that is liberated, some strict requirements have been set for the construction and design of the apparatus to be used (ASTM D-1756). The apparatus must contain an airflow meter and purifying train, a reaction unit fitted with a separator funnel and water-cooled condenser, a unit for removing interfering gases, and an absorber. The air-purifying train removes all carbon dioxide, and the water-cooled condenser removes moisture before it can enter the absorption train. Acid-forming gases, such as sulfur dioxide and hydrogen sulfide, and halogen acids are produced in the reaction and must be removed before entering the CO_2 absorber. Otherwise, they will be weighed as absorbed and measured as carbon dioxide. Anhydrous copper sulfate on pumice or granular silver sulfate is positioned in the absorption train to remove these interfering gases from the airstream before it enters the CO_2 absorber. The entire system must be gastight to prevent error, and a time schedule is specified to ensure repeatability and reproducibility.

Another option (ASTM D-6316) involved determination of the total, combustible, and carbonate carbon remaining in the solid by-products of combustion from boiler furnaces and similar reactors, including ash, fly ash, char, slag, and similar materials. The determination of total carbon is made by the oxidative thermal decomposition of a weighed quantity of sample in a closed system, and after complete oxidation and purification of the resulting gaseous products, measurement of the carbon dioxide produced by one of several methods. The evolved carbon dioxide is fixed on an absorption train and is measured quantitatively by weighing the absorbent (ASTM D-3178). The carbon dioxide is measured quantitatively by an electronic detection system calibrated against an appropriate reference standard (ASTM D-5373).

This test method is intended for the use of industry to determine the performance of boiler furnaces and similar combustion reactors and aid in determining the quality of the solid residue from combustion. Any of several methods can be used to determine the total carbon content combined with any of several methods to determine carbonate carbon, and the calculation, by difference, of the combustible carbon remaining in the sample. Alternatively, this test method applies to the determination of total carbon remaining in a material after acidification with strong acid to evolve carbonate carbon.

4.7.2 Arsenic and Selenium

Arsenic and selenium occur in coal to the extent of several parts per million and on combustion of the coal, varying quantities of these elements are released or retained in the ash, depending largely on the conditions under which the combustion takes place and on the nature of the coal ash.

Arsenic and selenium are determined (ASTM D-4606) by mixing a weighed sample with Eschka mixture and igniting at 750°C (1382°F). The mixture is dissolved in hydrochloric acid and the gaseous hydride of each element is generated from the appropriate oxidation state and determined by atomic absorption spectrophotometry. The method permits measurement of the total arsenic and selenium content of coal for the purpose of evaluating these elements where they can be of concern, for example, in coal combustion. When coal samples are prepared for analysis in accordance with standard procedure (ASTM D-2013), the arsenic and selenium are quantitatively retained and are representative of the total amounts in the coal.

REFERENCES

Ahmed, S. M., and Whalley, B. J. P. 1978. In *Analytical Methods for Coal and Coal Products*, Vol. 1, C. Karr, Jr. (Editor). Academic Press, San Diego, CA, Chap. 8.

ASTM. 2004. *Annual Book of ASTM Standards*, Vol. 05.06. American Society for Testing and Materials, West Conshohocken, PA. Specifically:

ASTM D-1756. Standard Test Method for Determination as Carbon Dioxide of Carbonate Carbon in Coal.

ASTM D-1757. Standard Test Method for Sulfate Sulfur in Ash from Coal and Coke.

ASTM D-2013. Standard Practice of Preparing Coal Samples for Analysis.

ASTM D-2015. Standard Test Method for Gross Calorific Value of Coal and Coke by the Adiabatic Bomb Calorimeter.

ASTM D-2361. Standard Test Method for Chlorine in Coal.

ASTM D-2492. Standard Test Method for Forms of Sulfur in Coal.

ASTM D-3173. Standard Test Method for Moisture in the Analysis Sample of Coal and Coke.

ASTM D-3174. Standard Test Method for Ash in the Analysis Sample of Coal and Coke from Coal.

ASTM D-3176. Standard Practice for Ultimate Analysis of Coal and Coke.

ASTM D-3177. Standard Test Methods for Total Sulfur in the Analysis Sample of Coal and Coke.

ASTM D-3178. Standard Test Methods for Carbon and Hydrogen in the Analysis Sample of Coal and Coke.

ASTM D-3179. Standard Test Methods for Nitrogen in the Analysis Sample of Coal and Coke.

ASTM D-3286. Standard Test Method for Gross Calorific Value of Coal and Coke by the Isoperibol Bomb Calorimeter.

ASTM D-3684. Standard Test Method for Total Mercury in Coal by the Oxygen Bomb Combustion/Atomic Absorption Method.

ASTM D-4208. Standard Test Method for Total Chlorine in Coal by the Oxygen Bomb Combustion/Ion Selective Electrode Method.

ASTM D-4239. Standard Test Methods for Sulfur in the Analysis Sample of Coal and Coke Using High Temperature Tube Furnace Combustion Methods.

ASTM D-4606. Standard Test Method for Determination of Arsenic and Selenium in Coal by the Hydride Generation/Atomic Absorption Method.

ASTM D-5373. Standard Test Methods for Instrumental Determination of Carbon, Hydrogen, and Nitrogen in Laboratory Samples of Coal and Coke.

ASTM D-6316. Standard Test Method for Determination of Total, Combustible and Carbonate Carbon in Solid Residues from Coal and Coke.

ASTM D-6414. Standard Test Method for Total Mercury in Coal and Coal Combustion Residues by Acid Extraction or Wet Oxidation/Cold Vapor Atomic Absorption.

ASTM D-6722. Standard Test Method for Total Mercury in Coal and Coal Combustion Residues by Direct Combustion Analysis.

ASTM E-256. Standard Test Method for Chlorine in Organic Compounds by Sodium Peroxide Bomb Ignition.

Attar, A. 1979. In *Analytical Methods for Coal and Coal Products*, Vol. 3, C. Karr, Jr. (Editor). Academic Press, San Diego, CA, Chap. 56.

BS. 2003. *Methods for Analysis and Testing of Coal and Coke*. BS 1016. British Standards Association, London.

Canfield, D. R., Ibarra, S., and McCoy, J. D. 1979. *Hydrocarbon Process.*, 58(7):203.

Chakrabarti, J. N. 1978a. In *Analytical Methods for Coal and Coal Products*, Vol. 1, C. Karr, Jr. (Editor). Academic Press, San Diego, CA, Chap. 9.

Chakrabarti, J. N. 1978b. In *Analytical Methods for Coal and Coal Products*, Vol. 1, C. Karr, Jr. (Editor). Academic Press, San Diego, CA, Chap. 10.

Chakrabarti, J. N. 1982. In *Coal and Coal Products: Analytical Characterization Techniques*, E. L. Fuller Jr. (Editor). Symposium Series 205. American Chemical Society, Washington, DC, Chap. 8.

Frigge, J. 1984. *Erdoel Kohle Erdgas Petrochem.*, 37:267.

Gluskoter, H. J., Shimp, N. F., and Ruch, R. R. 1981. In *Chemistry of Coal Utilization*, 2nd Suppl. Vol., M. A. Elliott (Editor). Wiley, Hoboken, NJ, Chap. 7.

Gorbaty, M. L., Kelemen, S. R., George, G. N., and Kwiatek, P. J. 1992. *Fuel*, 71:1255.

Hower, J. C., Riley, J. T., Thomas, G. A., and Griswold, T. B. 1992. *J. Coal Qual.*, 10:152.

DIN. 2004. *Standards for Fuel Analysis*. Deutsches Institut für Normung, Berlin. Specifically:

DIN 15722. Testing of Solid Fuels; Determination of Nitrogen Content; Dumas Method.

DIN 51722. Testing of Solid Fuels; Determination of Nitrogen Content; Semi-Micro Kjeldahl Method.

ISO. 1994. *Hard Coal—Determination of Oxygen Content*. International Organization for Standardization, Geneva, Switzerland.

ISO. 2003. *Standard Test Methods for Coal Analysis*. International Organization for Standardization, Geneva, Switzerland. Specifically:

ISO 157. Determination of Forms of Sulfur.

ISO 333. Determination of Nitrogen: Semi-micro Kjeldahl Method.

ISO 334. Determination of Total Sulfur: Eschka Method.

ISO 351. Determination of Total Sulfur: High Temperature Combustion Method.

ISO 352. Determination of Chlorine: High Temperature Combustion Method.

ISO 587. Determination of Chlorine Using Eschka Mixture.

ISO 602. Coal: Determination of Mineral Matter.

ISO 609. Determination of Carbon and Hydrogen: High Temperature Combustion Method.

ISO 625. 2003. Determination of Carbon and Hydrogen: Liebig Method.

ISO 15237. 2003. Determination of Total Mercury Content of Coal.

Kuhn, J. K. 1977. In *Coal Desulfurization*, T. D. Wheelock (Editor). Symposium Series 64. American Chemical Society, Washington, DC, p. 16.

Raymond, R., Jr. 1982. In *Coal and Coal Products: Analytical Characterization Techniques*, E. L. Fuller, Jr. (Editor). Symposium Series 205. American Chemical Society, Washington, DC, Chap. 9.

Slack, A. V. 1981. In *Chemistry of Coal Utilization*, 2nd Suppl. Vol., M. A. Elliott (Editor). Wiley, Hoboken, NJ, Chap. 22.

Volborth, A. 1979a. In *Analytical Methods for Coal and Coal Products*, Vol. 3, C. Karr, Jr. (Editor). Academic Press, San Diego, CA, Chap. 47.

Volborth, A. 1979b. In *Analytical Methods for Coal and Coal Products*, Vol. 3, C. Karr, Jr. (Editor). Academic Press, San Diego, CA, Chap. 55.

Volborth, A., Dahy, J. P., and Miller, G. E. 1987. In *Coal Science and Chemistry*, A. Volborth (Editor). Elsevier, Amsterdam, p. 417.

5 Mineral Matter

The term *mineral matter* refers to the inorganic constituents of coal and is all of the elements that are not part of the organic coal substance (carbon, hydrogen, nitrogen, oxygen, and sulfur) (Table 5.1). The mineral matter is the principal source of the elements that make up the ash when the coal is burned in air or oxygen. Four of the five elements generally considered to be organic (carbon, hydrogen, oxygen, and sulfur) are also present in inorganic combination in coals. Carbon is present in mineral (usually, calcium, magnesium, and iron) carbonates; hydrogen is present in free water and in water of hydration; oxygen is present in oxides, water, sulfates, and silicates; and sulfur is present in sulfides and sulfates.

Mineral matter in coal is usually classified as (1) *inherent mineral matter*, or (2) *adventitious mineral matter. Inherent mineral matter* is the inorganic material that is too closely associated with the coal substance to be readily separated from it by methods available at present. *Adventitious mineral matter* is the inorganic material that is less intimately associated with the coal and can readily be separated. There are also suggestions that the minerals transported and deposited in the peat swamp by wind and water be called *allogenic* or *detrital*, and that the remaining minerals, all of which formed in place (*authigenic*), be divided into those that formed contemporaneously with coal formation (*syngenetic*) and those whose formation followed the initial stages of coalification (*epigenetic*).

Mineral matter generally represents a significant proportion of coal composition, and the amount of mineral mater in coal varies from seam to seam, even along the same seam. Coals having mineral matter up to 32% by weight have been identified, and although a reasonable value for the *average* amount of mineral matter is much lower, caution is advised when using *average* numbers. The average usually bears no relationship to reality, where the range can vary from considerably below the average to considerably above the average. Coal performance on the basis of the average may be acceptable, but use of high-mineral-matter coal may cause considerable problems in a power plant. Generally, mineral matter in coal (whatever the content) is considered both undesirable and detrimental in coal utilization, and the presence of mineral matter affects almost every aspect of mining, preparation, transportation, and utilization.

Coal preparation is aimed at reducing the quantity of mineral matter, and efficient use of the methods chosen depends on its concentration and composition. However, no matter how effective the coal preparation technique, there is

Handbook of Coal Analysis, by James G. Speight
ISBN 0-471-52273-2

TABLE 5.1 Coal Minerals and Their Origins

Mineral Group	Syngenetic Formation (Intimately Intergrown): Transported by Water or Wind	Syngenetic Formation (Intimately Intergrown): Newly Formed	Epigenetic Formation: Deposited in Fissures, Cleats, and Cavities (Coarsely Intergrown)	Epigenetic Formation: Transformation of Syngenetic Minerals (Intimately Intergrown)
Clay minerals	Kaolinite, illite, sericite, clay minerals with mixed-layer structure, "Tonstein"		—	Illite, chlorite
Carbonates	—	Siderite–ankerite concentrations, dolomite, calcite, ankerite	Ankerite, calcite, dolomite	—
		Siderite, calcite, ankerite in fusite		
Sulfide ores	—	Pyrite concretions, melnikowite–pyrite, coarse pyrite (marcasite), concretions of FeS_2–$CuFeS_2$–ZnS	Pyrite, marcasite, zinc sulfide (sphalerite), lead sulfide (galena), copper sulfide (chalcopyrite)	Pyrite from the transformation of syngenetic concretions of $FeCO_3$
		Pyrite in fusite		
Oxide ores	—	Hematite	Goethite, lepidocrocite (needle iron ore)	—
Quartz	Quartz grains	Chalcedony and quartz from the weathering of feldspar and mica	Quartz	—
Phosphates	Apatite	Phosphorite	—	—
Heavy minerals and accessory minerals	Zircon, rutile, tourmaline, orthoclase, biotite	—	Chlorides, sulfates, nitrates	—

Source: Adapted from Murchison and Westoll (1968).

always a significant amount of residual mineral matter. This residual material is of considerable importance in coal utilization.

When coal is burned in a combustion unit, mineral matter undergoes major changes that lead to problems of clinker formation, fly ash, slagging, and boiler tube corrosion. The efficiency of a combustion unit is related to the amount of ash produced, since it is a diluent. On the positive side, ash has been utilized as a construction material and is a possible source of refractory materials, as cement additives, or as adsorbents for gas-cleaning processes. However, the composition of the ash must be known before it can be utilized in this way.

One of the principal impediments to the study of mineral matter in coal has been the difficulty of obtaining statistically valid (representative) samples of the mineral phases that are free of organic matter. At first glance it appears that it would be simple to separate minerals from coal by specific gravity techniques. Coal, the organic portion, has a low specific gravity (in the range 1.2 to 1.4), whereas the minerals occurring in coal have a specific gravity of 2 to 5 and greater. The finely disseminated nature of the minerals in coal precludes their complete separation, however. An incomplete separation based on specific gravity, which is the basis of many *coal cleaning* or *coal washing* processes, is effective in fractionating the sample into some portions that are relatively mineral rich and some that contain relatively pure coal. Specific gravity can be utilized to determine modes of occurrence of chemical elements and minerals in coal, and this makes the procedure important, but the procedure will probably not provide an acceptable representative sample of minerals in coal.

5.1 MINERAL TYPES

A large number of distinct mineral phases have been reported in various coals (Table 5.1), although lists of minerals in coal may contain as many as 50 to 60 minerals, most fall into one of five groups: (1) aluminosilicate minerals (clay minerals), (2) sulfide and sulfate minerals, (3) carbonate minerals, (4) silicate minerals (principally quartz), and (5) other minerals that include minerals that may occur in trace amounts or may be specific to a particular coal having originated because of the localized deposition and maturation conditions (Speight, 1994, and references cited therein).

5.1.1 Aluminosilicates (Clay Minerals)

Clay minerals are the most common inorganic constituents of coal and of the strata associated with coal seams. Many different clay minerals have been reported within and associated with coals, but the most common clay minerals are kaolinite and mixed-layer illite–montmorillonite. Kaolinite-rich clay is commonly found within and associated with coals in most of the coal basins of the world. They are generally called either *tonstein* or *kaolin-tonstein*.

5.1.2 Sulfide Minerals

The dimorphs pyrite (FeS_2) and marcasite (FeS_2) are the dominant sulfide minerals in coal; pyrite is the more abundant. Pyrite and marcasite have different crystal forms; pyrite is isometric and marcasite is orthorhombic. These two minerals are readily observed and, to some degree, easily removed as well as being especially interesting because they contribute significantly to the total sulfur content that causes boiler tube fouling, corrosion, and pollution by emission of sulfur dioxide when coal is burned (Beer et al., 1992).

5.1.3 Sulfate Minerals

The sulfate minerals identified in coal do not generally comprise a significant portion of the mineral matter in fresh, unoxidized coal samples. The iron disulfides oxidize rapidly after the coal is mined, however, and a number of hydrated sulfates ($FeSO_4 \cdot xH_2O$) have been reported in weathered coals and in coal refuse banks. The sulfates gypsum ($CaSO_4 \cdot 2H_2O$) and barite (Ba_2SO_4) are found in fresh coal. Most of the sulfates that form on weathering (oxidation) of pyrite are various hydrated states of ferrous and ferric sulfate.

5.1.4 Carbonate Minerals

The major cations found in the carbonate minerals in coals are calcium, magnesium, and iron. The rather pure end member calcite ($CaCO_3$) is dominant in some coals, whereas siderite ($FeCO_3$) is dominant in others. Calcite and ankerite (a mixed crystal composed of Ca, Mg, and Fe carbonates) are abundant in some coals.

5.1.5 Silicate Minerals

Quartz is the dominant form in which silica is found in coals, and it is ubiquitous. There is some distinction between clastic grains of quartz introduced by wind or water and authigenic quartz deposited from solutions. Quartz is also a major component of clay and siltstone partings in coal that are of detrital origin.

5.1.6 Other Minerals

A large number of minerals, in addition to those already discussed, have been reported to occur in coal. Not all have been positively identified, and often it is impossible to determine from the reports whether the mineral was intimately associated with the coal or was in the rock units making up the roof, floor, or a parting within the seam. Most of these other minerals are of limited significance in coal utilization, but a few are worth noting. Authigenic apatite [calcium fluorochlorohydroxyphosphate, $Ca_5(PO_4)_3 \cdot F \cdot Cl \cdot OH$] has been found in coal produced in widely separated areas of the world.

5.2 CHEMISTRY OF ASH FORMATION

When coal burns in air, as in the determination of ash in proximate analysis, all the organic material is oxidized or decomposed to give volatile products, and the inorganic material associated with the coal is subjected to the combined effects of thermal decomposition and oxidation. As a result, the quantity and composition of the resulting ash differ considerably from those of the inorganic materials originally associated with the pure coal substance.

It is therefore impossible to determine accurately the composition of the pure coal substance from the usual ultimate analysis simply by making allowance for the quantity of ash left behind as a residue when the coal is burned. Results obtained in this fashion are, as a consequence, quoted as being on a *dry, ash-free basis*, and no claim is therefore made that these results do in fact represent the composition of the pure coal substance. If, however, it were possible to calculate accurately the quantity of mineral matter originally present in the coal sample, then by making due allowance for this material, the composition of the pure coal material could be deduced with reasonable precision and certainly with a greater accuracy than could be obtained by adopting the analytical figures calculated to a dry, ash-free basis.

The thermal and oxidation changes in which the mineral matter takes part during the combustion of coal are very complex, but fortunately, the majority of the reactions are fairly clearly understood (Table 5.2) and it has been possible to derive a formula that enables the mineral matter content of the original coal to be calculated from a knowledge of the quantity of ash produced on combustion, together with the quantities of pyrite sulfur, chlorine, and carbonate present in the coal. The formula has been deduced from a consideration of the stoichiometric relationships existing between the reactants and products in reactions of the following types, all of which take place during combustion of coal and which together account for practically all the inorganic reactions involved in the process:

TABLE 5.2 Reactions of Coal Minerals at High Temperature

Inorganic Species	Behavior on Heating
Clays	Loose structural OH groups with rearrangements of structure and release of H_2O
Carbonates	Decompose with loss of CO_2; residual oxides fix some organic and pyritic S as sulfate
Quartz	Possible reaction with iron oxides from pyrite and organically held Ca in lignites; otherwise, no reaction
Pyrite	In air, burns to Fe_2O_3 and SO_2; in volatile matter test, decomposes to FeS
Metal oxides	May react with silicates
Metal carboxylates (lignites and subbituminous only)	Decompose; carbon in carboxylate may be retained in residue

It is difficult to determine, either qualitatively or quantitatively, the mineral matter content of a coal from the high-temperature ash. During high-temperature ashing, as designated by various standards [usually 750°C (1382°F)], a series of reactions takes place involving the minerals in the coals. Of the major mineral groups, only quartz is not significantly altered during high-temperature ashing.

The clay minerals in coal contain water that is bound within their lattices. Kaolinite contains 13% bound water, illite contains 4.5% bound water, and montmorillonite contains 5% bound water. In addition, montmorillonite that occurs in mixed-layer clays also contains interlayer or adsorbed water. All the water is lost during high-temperature ashing,

$$Al_2O_3 \cdot 2SiO_2 \cdot xH_2O \rightarrow Al_2O_3 \cdot 2SiO_2 + xH_2O \quad (5.1)$$

During high-temperature ashing, pyrite and marcasite (FeS_2) are oxidized to ferric sulfate [$Fe_2(SO_4)_3$] and sulfur dioxide (SO_2). Some of the sulfur dioxide may remain in the ash in combination with calcium, but much is lost. If all of the possible sulfur dioxide is emitted during ashing, there would be a 33% loss in weight with respect to the weight of pyrite or of marcasite in the original sample:

$$4FeS_2 + 11O_2 \rightarrow 2Fe_2O_3 + 8SO_2 \quad (5.2)$$

Calcium carbonate (calcite, $CaCO_3$) is calcined to lime (CaO) during high-temperature ashing, and carbon dioxide is evolved, resulting in a 44% reduction in weight from the original calcite:

$$CaCO_3 \rightarrow CaO + CO_2 \quad (5.3)$$

Other metal carbonates behave similarly (i.e., the oxides are formed during the ashing procedure). The stable mineral quartz (silicon dioxide, silica, SiO_2) is the only major mineral found in coal that is inert during high-temperature ashing.

Sulfates in coal ash are derived from two sources: (1) sulfates (generally of calcium or magnesium) present as such in the coal sample, and (2) sulfates formed by the absorption of sulfur oxides by the basic constituents of the ash during incineration of the coal and occurring mainly as calcium sulfate but also, to some extent, as alkali sulfates. The quantity of sulfates from the first source is usually very small and, in normal cases, can generally be neglected in comparison with those derived by absorption. Consequently, a determination of the sulfate in the ash enables a measure of the quantity formed during incineration to be established.

During combustion of the coal, organic chlorides are decomposed and liberate the chlorine atom as hydrochloric acid, while inorganic chlorides decompose with evolution of hydrochloric acid and ultimately leave a residue of the metallic oxide. Since approximately half the chlorine in coal occurs as inorganic chloride and the remainder as inorganic chloride, a correction for the chlorine present originally as inorganic material is easily applied when the total chlorine is known.

Except for oxygen and sulfur, elements that normally constitute the ash residues derived from coal combustion can arbitrarily be grouped as follows: (1) *major elements* (i.e., elements in concentrations greater than 0.5% in the whole coal, and these normally include aluminum, calcium, iron, and silicon), (2) *minor elements* (i.e., those in the range of concentration of about 0.02 to 0.5% in the whole coal, and these usually include potassium, magnesium, sodium, and titanium, and sometimes phosphorus, barium, strontium, boron, and others, depending on the geologic area), and (3) *trace elements* [i.e., all other inorganic elements usually detected in coal at less than 0.02% (200 ppm) down to parts per billion and below]. Most nonmetallic elements, even though they are more volatile than metals, leave a detectable residue in coal ash.

5.3 TEST METHOD PROTOCOLS

The determination of the ash content of coal is an essential part of coal evaluation (Shipley, 1962; Rees, 1966; Given and Yarzab, 1978; Weaver, 1978; Huggins et al., 1982; Nadkarni, 1982). However, ash, as stated above, is the residue remaining after burning the coal or coke under rigidly controlled conditions of sample weight, temperature, time, atmosphere, and equipment specifications. Thus, the mineral matter content of coal cannot be determined qualitatively or quantitatively from the ash that is formed when the coal is oxidized. The high-temperature ashing of coal at 750°C (1382°F) (ASTM D-3174; ISO 1171) causes a series of reactions involving the minerals in coal (see above).

Ash, as determined by the standard test method (ASTM D-3174), is the residue remaining after burning the coal and coke and differs in composition from the original inorganic constituents present in the coal. Incineration causes an expulsion of all water, the loss of carbon dioxide from carbonates, the conversion of iron pyrites into ferric oxide, and other chemical reactions. In addition, the ash, as determined by this test method, will differ in amount from ash produced in furnace operations and other firing systems because incineration conditions influence the chemistry and amount of the ash.

Approximately 1 g of the sample (weight to the nearest 0.1 mg) that has been pulverized to pass through a No. 60 sieve (ASTM D-2013) is placed in a cold muffle furnace and heated at such a rate that the temperature reaches 450 to 500°C (842 to 932°F) after 1 hour. The heating is continued so that a final temperature of 700 to 750°C (1292 to 1382°F) is reached by the end of the second hour, and heating is continued at this temperature for an additional 2 hours. At the end of the heating period, the sample is removed from the furnace, covered, allowed to cool under conditions that minimize moisture pickup, and weighed. The yield of ash is determined by the formula

$$\text{ash (\%w/w)} = \left(\frac{A - B}{C}\right) \times 100 \qquad (5.4)$$

where A is the weight of the sample container, cover, and ash residue; B the weight of the empty sample container and cover; and C the weight of the sample.

Although the 4-hour incineration period is often sufficient for most coals to reach a condition of complete burn-off, certain nonreactive coals may require additional time. If unburned carbon is observed in the sample container or if duplicate results are suspect, the samples should be returned to the furnace for sufficient time to reach a constant weight. On the other hand, there are suggestions that the 4-hour time limit may be reduced if the sample reaches a constant weight in less time. Caution is advised in using a reduced heating period and adherence to the 4-hour time period is recommended.

Some coal samples contain a high amount of carbonates (calcite, $CaCO_3$) or pyrite (FeS_2) or both. In such cases, sulfur retained as sulfates may be both unduly high and nonuniform between duplicate samples. Sulfate sulfur in the ash can be determined (ASTM D-1757) and the requisite correction made, and the ash yield should be reported and designated both as determined and corrected.

A reliable method of measuring the mineral matter content of a coal is an acid demineralization procedure. The method depends on the loss of weight of a sample when treated with 40% hydrofluoric acid at 50 to 60°C (122 to 140°F). Treatment of the sample with hydrochloric acid before and after treatment with hydrofluoric acid helps prevent the retention of insoluble calcium fluoride (CaF_2) in the coal. Pyrite is not dissolved in the treatment, consequently, pyrite and a small amount of retained chloride must be determined separately. Since two-thirds of the mass of the pyrite (FeS_2) is accounted for by the presence of ferric oxide (Fe_2O_3) in the residual ash, the mineral matter content is then given by the formula

$$\text{MM} = \text{weight loss} + \text{HCl} + (\text{FeS}_2) + \text{residual ash} \tag{5.5}$$

5.3.1 Data Handling and Interpretation

Contrary to popular belief and nomenclature, mineral matter and ash are not the same. Ash is produced by combustion of coal in oxygen, and the ash constituents, usually metal oxides, are formed from the mineral matter by routes involving some of the chemistry outlined above.

Determination of a good value for the percent of mineral matter content (% MM) is a very important component of coal analysis. If this quantity cannot be determined directly by the acid demineralization or low-temperature ashing procedure discussed previously, or by other suitable methods, it is possible to calculate a reasonable value for the mineral matter in coal, provided that the necessary data are available.

Because of the changes that occur in mineral matter during the ashing procedure, a number of formulas have been devised to allow calculation of the mineral matter content from various parameters determined during analyses of the coal. Minimally, this involves a calculation based on total content of ash and sulfur. The two formulas most commonly used are the Parr formula and the King–Maries–Crossley formula.

The Parr formula is the one most often used in the United States and requires only ash and sulfur values as determined in routine analysis:

$$\%MM = 1.08A + 0.55S \quad (5.6)$$

where A is the percentage of ash and S is the percentage of sulfur. The first term in this formula, 1.08A, is a correction for the loss in weight due to the elimination of water in the decomposition of clay minerals at high temperatures. Water of hydration of mineral matter has been estimated to be 8% of the ash value. The second term in the formula is a correction for the loss in weight when pyrite burns to ferric oxide (Fe_2O_3). The Parr formula treats all sulfur as pyritic and makes no allowance for the decomposition of carbonates or fixation of sulfur in the ash.

The King–Maries–Crossley formula is a more elaborate formula that allows for a number of effects. The original formula is

$$\%MM = 1.09A + 0.5S_{pyr} + 0.8CO_2 - 1.1SO_{3ash} + SO_{3coal} + 0.5Cl \quad (5.7)$$

where A is the percentage of ash, S_{pyr} the percentage of pyritic sulfur, CO_2 the percentage of mineral carbon dioxide, SO_{3ash} the percentage of SO_3 in ash, SO_{3coal} the total sulfur appearing as sulfates in coal, and Cl the percentage of chlorine. In this formula the various numbers represent correction factors for the loss in weight due to the elimination of water in the decomposition of clay minerals (1.09), for the oxidation of pyrite to ferric oxide and sulfur dioxide (0.5), for the loss of carbon dioxide from mineral carbonates (0.8), and for the fixation of sulfur in the ash (1.1). The addition of the value representing the sulfate content of the coal sample and one-half the chlorine (assuming one-half the chlorine in coal is found in the mineral matter) completes the formula. The revised King–Maries–Crossley formula is

$$\%MM = 1.13A + 0.5S_{pyr} + 0.8CO_2 - 2.8S_{ash} + 2.8S_{coal} + 0.5Cl \quad (5.8)$$

With this formula, a reasonably accurate value of the mineral matter can be calculated, but many parameters need to be determined in order to perform the computation.

An ultimate analysis that can claim to represent the composition of the organic substance of a coal is said to be on a *dry, mineral-matter-free* (dmmf) basis. The dmmf basis is a hypothetical condition corresponding to the concept of a pure coal substance. Since the dry, ash-free basis for coal neglects changes in mineral matter when coal is burned, the dmmf basis is preferred whenever the mineral matter can be determined or calculated.

The classification of coal (ASTM D-388) depends on calculation of the volatile matter yield and fixed carbon values on a dmmf basis. Calorific values are calculated on a *moist, mineral-matter-free basis.* The Parr formula is used in the classification system to calculate the mineral matter from ash and sulfur data.

Finally, it should be noted that it is strictly incorrect to refer to the *ash content* of coal, and it is only correct to refer to the *mineral matter content* of coal.

5.3.2 Ash Analysis

The evaluation of coal mineral matter by the ashing technique can be taken further insofar as attempts can then be made to determine the individual metal constituents of the ash. On the occasion when the mineral matter has been separated from the coal successfully, it is then possible to apply any one of several techniques (such as x-ray diffraction, x-ray fluorescence, scanning electron microscopy; and electron probe microanalysis) not only to investigate the major metallic elements in coal but also to investigate directly the nature (and amount) of the trace elements in the coal (Jenkins and Walker, 1978; Prather et al., 1979; Raymond and Gooley, 1979; Russell and Rimmer, 1979; Jones et al., 1992). Generally, no single method yields a complete analysis of the mineral matter in coal and it is often necessary to employ a combination of methods.

The wet chemical or *classical* procedures, including many colorimetric and some electrochemical procedures, can be quite time consuming. They are generally very accurate, however; they can be precise, and they do not require special or expensive equipment or facilities. However, these methods are less sensitive than some instrumental methods.

One issue that has already been mentioned is the amount of sulfur in the ash that is due to a high amount of carbonates (calcite, $CaCO_3$), or pyrite (FeS_2), or both, in the coal. Sulfur retained as sulfates may be both unduly high and nonuniform between duplicate samples. The reasons vary from inconsistencies in the furnace temperature and furnace ventilation that have an influence on sulfur trioxide retention in the ash. Consequently, sulfur in ash as determined in the laboratory cannot be assumed to be equivalent to sulfur present in the mineral matter in coal or to the retention of sulfur in ash produced under the conditions of commercial utilization.

Sulfate sulfur in ash is determined (ASTM D-1757) and the requisite correction made, and the ash yield should be reported and designated both as determined and corrected. The sulfate sulfur so determined can be used to calculate the sulfur trioxide portion of ash so that the ash content or ash composition can be reported on a sulfur trioxide–free basis.

In this method (ASTM D-1757), a specified quantity of ash is digested in boiling dilute hydrochloric acid solution to which bromine water is added to convert sulfite that may be present to the sulfate form. After neutralization and precipitation of iron with ammonium hydroxide, the mixture is filtered, and sulfate in the filtrate is determined gravimetrically as barium sulfate ($BaSO_4$). Alternatively, a specified quantity of ash and Eschka mixture are ignited together in air. The sulfates are dissolved in hot water and separated from undissolved ash residue and magnesium oxide by filtration. Sulfate in the filtrate is determined gravimetrically as barium sulfate ($BaSO_4$). However, barium in coal ashes can result in incomplete recovery of sulfate sulfur. High iron content in coal can introduce error if during the sulfate precipitation the iron is either partially adsorbed by the barium sulfate precipitate or is coprecipitated as iron sulfate.

A compositional analysis of the ash in coal is often useful in the total description of the quality of the coal. Knowledge of ash composition is also useful in predicting the behavior of ashes and slags in combustion chambers. Utilization of the ash by-products of coal combustion sometimes depends on the chemical composition of the ash. In addition, concern over release of certain trace elements to the environment as a result of coal utilization has made determination of these elements an increasingly important aspect of coal analysis.

Major and minor elements in coal, having concentrations easily detectable by most modern analytical techniques, can be determined by a number of acceptable procedures. Various approaches, combining a. number of specific procedures, are frequently referenced in the literature. For example, the presently accepted procedure (ASTM D-2795) determines silicon, aluminum, iron, titanium, and phosphorus colorimetrically, calcium and magnesium chelatometrically, and sodium and potassium by flame photometry. This standard test method was withdrawn in 2001 but is still used in some laboratories.

Another test method (ASTM D-3682) covers analysis of the commonly determined major and minor elements (such as silicon, aluminum, iron, calcium, magnesium, sodium, potassium, and titanium) in laboratory coal ash and in combustion residues from coal utilization processes by atomic absorption/emission spectroscopy. In the test method, the ash sample or combustion residue to be analyzed is standardized by ignition in air at 750°C (1382°F) to a constant weight. The ash is fused within lithium tetraborate ($Li_2B_4O_7$) followed by a final dissolution of the melt in either dilute hydrochloric acid (HCl) or dilute nitric acid (HNO_3). The solution is analyzed by atomic absorption/emission spectroscopy for applicable elements. As always, the chemical composition of laboratory-prepared ash does not represent the composition of mineral matter in the coal or the composition of fly ash and slag resulting from commercial-scale burning of the coal, but it is recommended that sulfur be determined by other test methods (ASTM D-1757; ASTM D-5016).

The less common elements in coal ash (e.g., beryllium, chromium, copper, manganese, nickel, lead, vanadium, zinc, and cadmium) can also be determined using atomic absorption (ASTM D-3683). In the test method, the ash is dissolved by mineral acids, and the individual elements determined by atomic absorption spectrometry.

For certain analytical purposes, it may be desirable to separate the minerals from the coal in an unaltered form. In early studies, density separation methods were used, which were unsatisfactory because of the enrichment of certain minerals in the process. A low-temperature ashing, or plasma ashing, technique has been developed that is more reliable and faster than density separations (Gluskoter, 1965). In this method, low-pressure oxygen is activated by a radio-frequency discharge. The excited oxygen atoms and other oxygen-containing species oxidize the carbonaceous material at low temperatures (approximately 150°C, 302°F). The effects of low-temperature ashing and of the oxidizing gas stream on the minerals in coal are minimal. Some pyrite can be oxidized, and to some extent organic sulfur can be fixed as sulfates. The rates of these reactions are functions of operating conditions such as radio-frequency power level and oxygen flow rate.

Factors that affect the rate of low-temperature ashing other than radio-frequency power and oxygen flow rate are the coal particle size and depth of sample bed. Typical conditions for ashing are a particle size of less than 80 mesh, a sample layer density of 30 mg/cm^2, oxygen flow rate of 100 cm^3/min, chamber pressure of about 2 torr, and a 50-W net radio-frequency power. The total time required is 36 to 72 hours, and specified conditions must be met during the procedure to obtain reproducible results.

X-ray fluorescence analysis (ASTM D-4326) is a rapid, simple, and reasonably accurate method of determining the concentration of many minor and trace elements in whole coal. The method is dependent on the availability of suitable standards. Although the major elements in coal (carbon, hydrogen, oxygen, and nitrogen) cannot be analyzed by x-ray fluorescence, most other elements at levels greater than a few parts per million (ppm) are readily determined. Sulfur should be determined by an alternative method (ASTM D-1757).

In the test method (ASTM D-4326), the coal to be analyzed is ashed under standard conditions and ignited to constant weight. Previously ashed materials are ignited to constant weight under standard conditions. The ash is fused with lithium tetraborate ($Li_2B_4O_7$) or other suitable flux and either ground and pressed into a pellet or cast into a glass disk. The pellet or disk is then irradiated by an x-ray beam of short wavelength (high energy). The characteristic x-rays of the atom that are emitted or fluoresced upon absorption of the primary or incident x-rays are dispersed and intensities at selected wavelengths are measured by sensitive detectors. Detector output is related to concentration by calibration curves or by computerized data-handling equipment. All of the elements are determined as the element and reported as the oxide and include silicon, aluminum, iron, calcium, magnesium, sodium, potassium, phosphorus, titanium, manganese, strontium, and barium.

If the sample to be analyzed has been stored and the absorption of moisture or carbon dioxide, or both, is in question, the ash should be reignited using the 500 to 750°C (932 to 1382°F) staged combustion procedure before analysis. Alternatively, the ignition loss can be determined using the 500 to 750°C (932 to 1382°F) staged combustion procedure on a separate sample weighed out at the same time as the test portion and the necessary corrections made. Materials previously ashed, fly ash, or bottom ash must be ignited to constant weight at 750°C (1382°F) and cooled in a desiccator before analysis sample preparation, or alternatively, weight loss or gain must be determined on a second sample at 750°C (1382°F) taken at the same time as the analysis sample so that analysis determined on an as-received basis can be reported on an ignited ash basis.

Infrared absorption bands from 650 to 200 cm^{-1} have been used extensively to analyze, both qualitatively and quantitatively, minerals in coal ash. One such test method (ASTM D-5016) is a procedure using a high-temperature tube furnace and infrared detection for the rapid determination of sulfur in ash from coal and coke and is an alternative test method for a test method described previously (ASTM D-1757). The purpose of this test method is to determine the percent sulfur trioxide (SO_3) portion of the major and minor elements in coal ash. This test method may be used to determine the percent sulfur trioxide (SO_3) portion of ash (ASTM D-3174; ASTM D-5142) for coal containing high amounts of calcium. The ash may then be reported on a sulfur trioxide–free basis.

In the test method, a weighed test portion is mixed with a promoting agent (that aids in the quantitative release of all sulfur present in the test portion as sulfur dioxide) and ignited in a tube furnace at a minimum operating temperature of 1350°C (2460°F) in a stream of oxygen. Some promoters may contain sulfur, and the sulfur content should be determined by analyzing the promoter as a sample and appropriate corrections made based on the mass of the promoter and its sulfur content.

The combustible sulfur contained in the test portion is oxidized to gaseous oxides of sulfur. Moisture and particulates are removed by traps filled with anhydrous magnesium perchlorate. The gas stream is passed through a cell in which sulfur dioxide is measured by an infrared absorption detector. Sulfur dioxide absorbs infrared energy at a precise wavelength within the infrared spectrum. Energy is absorbed as the gas passes through the cell body in which the infrared energy is being transmitted; thus, at the detector, less energy is received. All other infrared energy is eliminated from reaching the detector by a precise wavelength filter. The absorption of infrared energy can be attributed only to sulfur dioxide, whose concentration is proportional to the change in energy at the detector. One cell is used as both a reference and a measurement chamber. Total sulfur as sulfur dioxide is detected on a continuous basis.

Ashing temperature, heating rate, and furnace ventilation have an important influence on sulfur trioxide retention; thus, observance of the prescribed ashing conditions can be critical. Sulfur in ash as determined by these methods cannot be strictly related to the sulfur oxides retained in ash produced under the conditions of combustion in boiler furnaces or other commercial combustion processes.

There is also a standard test method for determination of major and minor elements in coal ash by inductively coupled plasma (ICP)–atomic emission spectrometry (ASTM D-6349). In the test method, the sample to be analyzed is ashed under standard conditions and ignited to constant weight. The ash is fused with a fluxing agent followed by dissolution of the melt in dilute acid solution. Alternatively, the ash is digested in a mixture of hydrofluoric, nitric, and hydrochloric acids. The solution is analyzed by (ICP)–atomic emission spectrometry for the elements. The basis of the method is the measurement of atomic emissions. Aqueous solutions of the samples are nebulized, and a portion of the aerosol that is produced is transported to the plasma torch, where excitation and emission occurs. Characteristic line emission spectra are produced by a radio-frequency inductively coupled plasma. A grating monochromator system is used to separate the emission lines, and the intensities of the lines are monitored by photomultiplier tube or photodiode array detection. The photocurrents from the detector are processed and controlled by a computer system. A background correction technique is required to compensate for variable background contribution to the determination of elements. Background must be measured adjacent to analyte lines of samples during analysis. The position selected for the background intensity measurement, on either or both sides of the analytical line, will be determined by the complexity of the spectrum adjacent to the analyte line. The position used must be free of spectral interference and reflect the same change in background intensity as occurs at the analyte wavelength measured.

Several types of interference effects may contribute to inaccuracies in the determination of major and minor elements. The interferences can be classified as spectral, physical, and chemical. *Spectral interferences* involve an overlap of a spectral line from another element, unresolved overlap of molecular band spectra, background contribution from continuous or recombination phenomena, and background contribution from stray light from the line emission of high-concentration elements. The second effect may require selection of an alternative wavelength. The third and fourth effects can usually be compensated by a background correction adjacent to the analyte line.

Physical interferences are generally considered to be effects associated with such properties as change in viscosity, and surface tension can cause significant inaccuracies, especially in samples that may contain high dissolved solids, or acid concentrations, or both. If these types of interferences are operative, they must be reduced by dilution of the sample or utilization of standard addition techniques, or both.

Chemical interferences, which can be highly dependent on matrix type and the specific analyte element, are characterized by molecular compound formation, ionization effects, and solute vaporization effects. If such effects are observed, they can be minimized by careful selection of operating conditions, by buffering the sample, by matrix matching, and by standard addition procedures.

Coal contains several elements whose individual concentrations are generally less than 0.01%. These elements are commonly and collectively referred to as *trace elements*. These elements occur primarily as part of the mineral matter in coal. Hence, there is another standard test method for determination of major and minor elements in coal ash by ICP–atomic emission spectrometry, inductively coupled plasma mass spectrometry, and graphite furnace atomic absorption spectrometry (ASTM D-6357). The test methods pertain to the determination of antimony, arsenic, beryllium, cadmium, chromium, cobalt, copper, lead, manganese, molybdenum, nickel, vanadium, and zinc (as well as other trace elements) in coal ash.

In the test method, the coal or coke to be analyzed is ashed under controlled conditions, digested by a mixture of aqua regia and hydrofluoric acid, and finally dissolved in 1% nitric acid. The concentration of individual trace elements is determined by either inductively coupled plasma-atomic emission spectrometry (ICPAES) or inductively coupled plasma-mass spectrometry (ICPMS). Selected elements that occur at concentrations below the detection limits of ICPAES can be analyzed quantitatively by graphite furnace atomic absorption spectrometry (GFAA).

Use of a high-speed pulverizer for size reduction should be avoided. The use of jaw crushers followed by final preparation in an agate mortar and pestle is recommended to prevent contamination of the sample. Separate test portions should be analyzed for moisture content (ASTM D-3174; ASTM D-5142) so that calculations to other bases can be made.

An internal standard is needed to compensate for differences in physical properties (such as viscosity) between the calibration standard and the test samples and drift caused by thermal changes in the laboratory that will affect the instrument optics. An appropriate internal standard element should not be naturally present in the test samples in appreciable concentrations and should not present spectral interferences with any analyte. In addition, the internal standard should be a strong emitter so that its relative concentration can be kept low, and be as chemically similar to the analyte as possible.

Three methods for determining mineral carbon dioxide in coal were investigated using bituminous coal. The titrimetric method is claimed to be superior to either of the then-used British standard gravimetric or manometric methods (BS 1016). The procedure involves the decomposition of carbonate minerals with hydrochloric acid and absorption of the evolved carbon dioxide in a mixture of benzylamine, ethanol, and dioxan. This mixture forms a stable salt of benzylcarbamic acid, which is then titrated with sodium methoxide. The method was said to be suitable for all concentrations of carbon dioxide. It is especially accurate for low concentrations, and it is much more rapid than other methods tested.

Several other methods have been used to determine the trace elements in the mineral matter of coal, as well as in whole coal and coal-derived materials. These methods include spark-source mass spectrometry, neutron activation analysis, optical emission spectroscopy, and atomic absorption spectroscopy.

Spark-source mass spectrometry (SSMS) has been used extensively in the determination of trace elements in coal. Whole coal samples as well as ash residues, fly ash, and coal dust have been analyzed using this technique.

Neutron activation analysis techniques are frequently used for trace element analyses of coal and coal-related materials (Weaver, 1978). Precision of the method is 25%, based on all elements reported in coal and other sample matrices. Overall accuracy is estimated at 50%. Neutron activation analysis utilizing radiochemical separations (NAA-RC) is employed by investigators when the sensitivity for a particular element or group of elements is inherently low or when spectral interference for a given element in a specific matrix is too great to be detected adequately. This situation was more prevalent before the advent of Ge(Li) spectrometry when only low-resolution NaI(TI) detectors were available.

Although thermal (slow) neutrons derived from nuclear reactors are the most practical source of particles for nuclear excitation and generally provide the more useful reaction, other excitation sources, such as 14-MeV (fast) neutrons from commercially available accelerators or generators, have also been applied to coal analysis.

Optical emission spectroscopy has been extensively applied to the quantitative and semiquantitative determination of multiple trace elements in coal ash with a variety of excitation and optical parameters, internal standards, calibration standards, and procedures.

X-ray diffraction analysis is probably the best developed and most widely used method for identifying mineral phases in coal and is the most promising method for the identification of minerals in coals (Rao and Gluskoter, 1973). These techniques may be applied to minerals from coal after the mineral matter has been separated from organic matter in the coal. However, its application can be limited because of orientation effects, and a reliable method of sample preparation is necessary to prevent these from occurring. X-ray diffraction profiles are determined by using a conventional diffractometer system with monochromatic x-radiation. For qualitative analysis, the specimen is scanned over a wide angular range to ensure that all the major diffraction peaks of the component minerals are recorded. Diffraction spacings are then calculated from the peak positions, and the elements present in the sample are determined by using standard tables of diffraction spacing.

Scanning electron microscopy with an energy-dispersive x-ray system accessory has been used to identify the composition and nature of minerals in coals and to determine the associations of minerals with each other. Examinations can be made on samples resulting from ashing techniques or whole coal. With this technique it is possible to identify the elemental components and deduce the mineral types present in coal samples. Computerized systems to evaluate scanning electron microscopy images have been developed and are useful in characterizing the minerals in coal mine dusts and in coal.

Recent studies on iron sulfide minerals in coals, minerals in coals, and in situ investigation of minerals in coal all used the scanning electron microscope (SEM) as the primary analytical tool. The ion microprobe mass analyzer (IMMA) is more sensitive than either the energy-dispersive x-ray spectrometer or the wavelength-dispersive x-ray spectrometer, both of which are used as accessories to an electron microscope.

Optical microscopy is another method that has been used to determine the distribution of minerals in coal. This method is based on the detailed microscopic examination of polished or thin sections of coal in transmitted and/or reflected light. In principle, observing several of its optical properties, such as morphology, reflectance, refractive index, and anisotropy, makes identification of a mineral type possible.

REFERENCES

ASTM. 2004. *Annual Book of ASTM Standards*, Vol. 05.06. American Society for Testing and Materials, West Conshohocken, PA. Specifically:

ASTM D-388. Standard Classification of Coals by Rank.

ASTM D-1757. Standard Test Method for Sulfate Sulfur in Ash from Coal and Coke.

ASTM D-2013. Standard Practice of Preparing Coal Samples for Analysis.

ASTM D-2795. Standard Test Methods for Analysis of Coal and Coke Ash.

ASTM D-3174. Standard Test Method for Ash in the Analysis Sample of Coal and Coke from Coal.

ASTM D-3682. Standard Test Method for Major and Minor Elements in Combustion Residues from Coal Utilization Processes.

ASTM D-3683. Standard Test Method for Trace Elements in Coal and Coke Ash by Atomic Absorption.

ASTM D-4326. Standard Test Method for Major and Minor Elements in Coal and Coke Ash by X-ray Fluorescence.

ASTM D-5016. Standard Test Method for Sulfur in Ash from Coal, Coke, and Residues from Coal Combustion Using High-Temperature Tube Furnace Combustion Method with Infrared Absorption.

ASTM D-5142. Standard Test Methods for Proximate Analysis of the Analysis Sample of Coal and Coke by Instrumental Procedures.

ASTM D-6349. Standard Test Method for Determination of Major and Minor Elements in Coal, Coke, and Solid Residues from Combustion of Coal and Coke by Inductively Coupled Plasma–Atomic Emission Spectrometry.

ASTM D-6357. Test Methods for Determination of Trace Elements in Coal, Coke, and Combustion Residues from Coal Utilization Processes by Inductively Coupled Plasma Atomic Emission, Inductively Coupled Plasma Mass, and Graphite Furnace Atomic Absorption Spectrometries.

Beer, J. M., Sarofim, A. F., and Barta, L. E. 1992. *J. Inst. Energy*, 65(462):55.

BS. 2003. *Methods for Analysis and Testing of Coal and Coke*. BS 1016. British Standards Institution, London.

Given, P. H., and Yarzab, R. F. 1978. In *Analytical Methods for Coal and Coal Products*, Vol. 2, C. Karr, Jr. (Editor). Academic Press, San Diego, CA, Chap. 20.

Gluskoter, H. J. 1965. *Fuel*, 44:285.

Huggins, F. E., Huffman, G. P., and Lee, R. J. 1982. In *Coal and Coal Products: Analytical Characterization Techniques*, E. L. Fuller, Jr. (Editor). Symposium Series 205. American Chemical Society, Washington, DC, Chap. 12.

ISO. 2003. *Standard Test Methods for Coal Analysis*. International Organization for Standardization, Geneva, Switzerland. Specifically:

ISO 1171. Determination of Ash.

Jenkins, R. G., and Walker, P. L., Jr. 1978. In *Analytical Methods for Coal and Coal Products*, Vol. 2, C. Karr, Jr. (Editor). Academic Press, San Diego, CA, Chap. 26.

Jones, M. L., Kalmanovitch, D. P., Steadman, E. N., Zygarlicke, F. J., and Benson, S. A. 1992. In *Advances in Coal Spectroscopy*, H. C. Meuzelaar (Editor). Plenum Press, New York, Chap. 1.

Murchison, D., and Westoll, T. S. (Editors). 1968. *Coal and Coal-Bearing Strata*. Elsevier, Amsterdam.

Nadkarni, R. A. 1982. In *Coal and Coal Products: Analytical Characterization Techniques*, E. L. Fuller, Jr. (Editor). Symposium Series 205. American Chemical Society, Washington, DC, Chap. 6.

Prather, J. W., Guin, J. A., and Tarrer, A. R. 1979. In *Analytical Methods for Coal and Coal Products*, Vol. 3, C. Karr, Jr. (Editor). Academic Press, San Diego, CA, Chap. 49.

Rao, C. P., and Gluskoter, H. J. 1973. *Occurrence and Distribution of Minerals in Illinois Coals*. Circular No. 476. Illinois State Geological Survey, Urbana, IL.

Raymond, R., Jr., and Gooley, R. 1979. In *Analytical Methods for Coal and Coal Products*, Vol. 3, C. Karr, Jr. (Editor). Academic Press, San Diego, CA, Chap. 48.

Rees, O. W. 1966. *Chemistry, Uses and Limitations of Coal Analyses*. Report of Investigations 220. Illinois State Geological Survey, Urbana, IL.

Russell, S. J., and Rimmer, S. M. 1979. In *Analytical Methods for Coal and Coal Products*, Vol. III, C. Karr, Jr. (Editor). Academic Press, San Diego, CA, Chap. 42.

Shipley, D. E. 1962. *Mon. Bull. Br. Coal Utilization Res. Assoc.*, 26:3.

Speight, J. G. 1994. *The Chemistry and Technology of Coal*, 2nd ed. Marcel Dekker, New York.

Ward, C. R. 1977. *Mineral Matter in the Springfield-Harrisburg (No. 5) Coal Member in the Illinois Basin*. Circular No. 498. Illinois State Geological Survey, Urbana, IL.

Weaver, J. N. 1978. In *Analytical Methods for Coal and Coal Products*, Vol. 1, C. Karr, Jr. (Editor). Academic Press, San Diego, CA, Chap. 11.

6 Physical and Electrical Properties

Just as coal evaluation can be achieved by the determination of several noteworthy properties (Chapter 3 and 4), there are various physical and electrical properties (Table 6.1) that provide even more valuable information about the potential use for coal (van Krevelen, 1957; Evans and Allardice, 1978). Indeed, there are also those properties of organic materials that offer valuable information about environmental behavior (Lyman et al., 1990): an additional reason to study the properties of coal.

Knowledge of the electrical properties of coal is also an important aspect of coal characterization and behavior. Electrical properties are useful for cleaning, mining, pyrolysis, and carbonizing processes. They are also of special interest in the electro-linking process for permeability enhancement and as a means to locate regions with different physical properties during in situ coal gasification. These properties and the behavior of coal play an important part in dictating the methods by which coal should be handled and utilized.

In the broadest sense, it has been suggested that the granular nature of high-rank coals is of importance in understanding the physical nature of coal if coal is modeled in terms of a granular medium that consists of graphitelike material embedded in an insulating organic matrix (Speight, 1994, and references cited therein). Indeed, there have been several earlier suggestions of the graphitelike nature of coal, particularly from x-ray diffraction studies (Speight, 1978, and references cited therein) and perhaps this is a means by which the behavior of coal can be modeled. But if this is the case, the precise role of the smaller aromatic systems needs also to be defined more fully. Nevertheless, it certainly offers new lines of thinking about coal behavior.

At first consideration, there may appear to be little, if any, relationship between the physical and chemical behavior of coal, but in fact the converse is very true. For example, the pore size of coal (which is truly a physical property) is a major factor in determining the chemical reactivity of coal (Walker, 1981). And chemical effects that result in the swelling and caking of coal(s) have a substantial effect on the means by which coal should be handled either prior to or during conversion operations. This is the reason for the study and measurement of these very important properties of coal.

Handbook of Coal Analysis, by James G. Speight
ISBN 0-471-52273-2

TABLE 6.1 Physical Properties Used for Determining Coal Suitability for Use

Test/Property	Results/Comments
Physical properties	
Density	True density as measured by helium displacement
Specific gravity	Apparent density
Pore structure	Specification of the porosity or ultrafine structure of coals and nature of pore structure between macro, micro, and transitional pores
Surface area	Determination of total surface area by heat of absorption
Reflectivity	Useful in petrographic analyses
Mechanical properties	
Strength	Specification of compressibility strength
Hardness/abrasiveness	Specification of scratch and indentation hardness; also abrasive action of coal
Friability	Ability to withstand degradation in size on handling, tendency toward breakage
Grindability	Relative amount of work needed to pulverize coal
Dustiness index	Amount of dust produced when coal is handled
Thermal properties	
Calorific value	Indication of energy content
Heat capacity	Measurement of the heat required to raise the temperature of a unit amount of coal 1°
Thermal conductivity	Time rate of heat transfer through unit area, unit thickness, unit temperature difference
Plastic/agglutinating	Changes in a coal upon heating; caking properties of coal
Agglomerating index	Grading on nature of residue from 1-g sample when heated at 950°C (1550°F)
Free swelling index	Measure of the increase in volume when a coal is heated without restriction
Electrical properties	
Electrical resistivity	Electrical resistivity of coal measured in ohm-centimeters
Dielectric constant	Measure of electrostatic polarizability

When determining the physical properties of coal, there is often considerable debate about the *correctness* of the results obtained from measurements by two or more different analytical techniques. Provided that the methods and/or equipment used was capable of producing high-quality data, the pertinent issues relate to whether or not the sample properly prepared and properly measured and whether or not the analytical parameters applied correctly in the data-handling step(s). Thus, the concept of *different techniques yielding different, albeit correct results* can apply to the measurement of physical properties such as density, porosity, particle size, and surface area.

6.1 DENSITY AND SPECIFIC GRAVITY

For porous solids such as coal, there are five different density measurements: *true density, apparent density, particle density, bulk density*, and *in-place density*. The *true density* of coal is the mass divided by the volume occupied by the actual, pore-free solid in coal. However, determining mass of coal may be deemed as being rather straightforward, but determining volume presents some difficulties. *Volume*, as the word pertains to a solid, cannot be expressed universally in a simple definition. Indeed, the method used to determine volume experimentally, and subsequently, the density, must be one that applies measurement rules consistent with the adopted definition.

The precise determination of true density requires complete filling of the pore structure with a fluid that has no interaction with the solid. No fluid meets these requirements completely. Helium has traditionally been considered as the best choice since it is not significantly adsorbed by coal at room temperature and that the use of helium gives a more accurate determination of coal density, but there is evidence (Berkowitz, 1979, and references cited therein) that part of the pore system may be inaccessible to the helium. Thus, when helium is used as the agent for determining coal density, the density (helium density) may differ from the true density and may actually be lower than the true density.

Thus, it is not surprising that coal density is variable and dependent on the coal type. For example, the density of anthracite is on the order of 1.55, whereas bituminous coal has a density on the order of 1.35, and lignite has a density on the order of 1.25. However, such generalizations are to be treated with caution since coal density is usually determined by displacement of a fluid, but because of the porous nature of coal and also because of physicochemical interactions, the density data observed vary with the particular fluids employed, and different fluids may have to be employed for different coal types (Agrawal, 1959; Mahajan and Walker, 1978).

A standard test method is available for determining the true density (i.e., true specific gravity) of coke (ASTM D-167) that with careful application can be applied to coal. The method actually describes the determination of apparent specific gravity and true specific gravity of lump coke larger than 25-mm (1-in.) size and calculating porosity from the specific gravity data. Apparent and true specific gravity as determined by this test method are influenced by the type of coals carbonized and the operating and preparational conditions of that carbonization, that is, charge bulk density, heating rate, and pulverization level.

The test method involves the use of a cage or basket constructed of 13-mm ($\frac{1}{2}$-in.)-square mesh screen wire cloth, it is necessary to have pieces that will remain in the cage when it is removed from the water. Because it is very difficult to collect a representative sample, care must be taken to select pieces representing the entire exposed area, if sampling must be done in this manner. In addition, it may be necessary to dry the sample before crushing or pulverizing during sample preparation (ASTM D-346). In all cases, care must be taken to select equipment that will not abrade and add unwanted impurities to the sample during size-reduction steps.

The *true density* of coal is usually determined by helium displacement and therefore is often referred to as the *helium density*. Helium is used because it has the ability to penetrate all the pores of a given sample of coal without (presumably) any chemical interaction. In the direct-pressure method, a known quantity of helium and a weighed sample of coal are introduced into an apparatus of known volume, whereupon the pressure of the helium at a given temperature allows calculation of the volume of the coal. In the indirect method, mercury is used to compensate for the helium displaced by the introduction of the coal.

The true density (in fact, the helium density) has been correlated with the elemental composition of coal. Thus,

$$\text{density}_{\text{He}} = 1.534 - 0.05196\text{H} + 0.007375\text{O} - 0.02472\text{N} + 0.003853\text{S} \quad (6.1)$$

where $\text{density}_{\text{He}}$ is the helium density (g/cm^3) and H, O, N, and S are the respective weight percent of the various elements (dmmf basis).

Coal density is a useful parameter not only for deducing the spatial structure of coal molecules, but the relationship between the density and porosity suggests that emphasis must be given to density and its determination. Porosity measurement, in turn, provides useful information on the technical behavior of coal toward its end use. Particle density is required for calculating the porosity of individual coal particles.

Methods of measurement of coal density include use of a gas pycnometer and particle density by mercury porosimetry. However, the difference in density values using different gases must be recognized since, for example, density values measured by nitrogen may be greater than those obtained when helium is used. Density measurement depends on adsorption of gas molecules, and differences (between nitrogen and helium) may be due to nitrogen adsorption on the coal surface.

A gas pycnometer operates by detecting the pressure change resulting from displacement of gas by a solid object. Expanding a quantity of gas at known pressure into an empty chamber and measuring the pressure establishes a baseline. Then a sample is placed in the chamber and the chamber is resealed. The same quantity of gas at the same pressure is again expanded into the sample chamber, and the pressure is measured. The difference in the two pressures combined with the known volume of the empty sample chamber allows the volume of the sample to be determined by way of the gas law. The accuracy and precision of the gas pycnometer in determining density is good, but the method relies greatly on the cleanliness of the sample material and purity of the analysis gas. The accessibility to small open pores is quite high. The volume measured by this method will be less (density greater) than that determined by mercury porosimetry or other liquid displacement methods when small, open pores are in the sample.

Mercury is a nonwetting liquid that must be forced to enter a pore by application of external pressure. The surface tension of mercury causes mercury to bridge the openings of pores, cracks, and crevices until sufficient pressure is applied to force entry. For example, at atmospheric pressure, mercury will resist entering pores smaller than about 6 μm in diameter. When an object is surrounded by mercury, the mercury forms a closely fitting liquid envelope around the object. How closely the mercury conforms to the surface features of the object depends on the pressure applied. As pressure increases, mercury enters smaller and smaller voids in the sample. At a pressure of 60,000 psi, mercury has been forced to enter pores of diameters down to 0.003 μm. This fills essentially all pore volume in most materials. Typically, the volume of mercury displaced at minimum pressure and that displaced at maximum pressure are used to determine bulk (or envelope) density and skeletal density, respectively. For powders, the total volume of the grains can be determined by subtracting the interparticle void volume from the bulk volume. A mercury porosimeter is more often used for determination of pore volume distribution by pore size determination of density. Thus, pore volume is often a by-product of a data set from porosimetry measurements.

The *apparent density* of coal is determined by immersing a weighed sample of coal in a liquid followed by accurate measurement of the liquid that is displaced (pycnometer method). For this procedure, the liquid should (1) wet the surface of the coal, (2) not absorb strongly to the coal surface, (3) not cause swelling, and (4) penetrate the pores of the coal.

It is difficult (if not impossible) to satisfy all of these conditions, as evidenced by the differing experimental data obtained with solvents such as water, methanol, carbon tetrachloride, benzene, and other fluids. Thus, there is always the need to specify the liquid employed for the determination of density by means of this (pycnometer) method. Furthermore, a period of 24 hours may be necessary for the determination because of the need for the liquid to penetrate the pore system of the coal to the maximum extent.

The *particle density* is the weight of a unit volume of solid, including the pores and cracks (Mahajan and Walker, 1978). The particle density can be determined by any of three methods: (1) mercury displacement (Gan et al., 1972); (2) gas flow (Ergun, 1951); and (3) silanization (Ettinger and Zhupakhina, 1960).

The density of coal shows a notable variation with rank for carbon content (Figure 6.1) and, in addition, the methanol density is generally higher than the helium density because of the contraction of adsorbed helium in the coal pores as well as by virtue of interactions between the coal and the methanol, which results in a combined volume that is notably less than the sum of the separate volumes. Similar behavior has been observed for the water density of coals having 80 to 84% w/w carbon.

Coal with more than 85% w/w carbon have usually been shown to exhibit a greater degree of hydrophobic character than the lower-rank coals, with the additional note that the water density may be substantially lower than the helium density; for the 80 to 84% carbon coals, there is generally little, if any, difference between the helium and water densities. However, the hydrophobicity of coal correlates better with the moisture content than with the carbon content and better with the moisture/carbon molar ratio than with the hydrogen/carbon or oxygen/carbon atomic ratios. Thus, it appears that there is a relationship

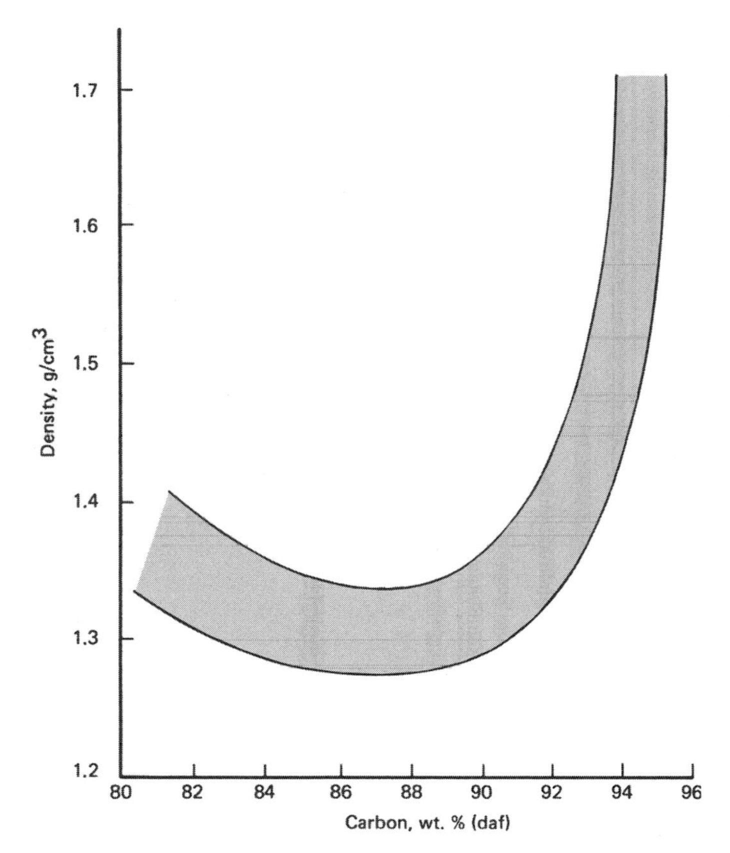

FIGURE 6.1 Variation of density with carbon content. (Adapted from Berkowitz, 1979.)

between the hydrophobicity of coal and the moisture content (Labuschagne, 1987; Labuschagne et al., 1988).

An additional noteworthy trend is the tendency for the density of coal to exhibit a minimum value at approximately 85% w/w carbon. For example, a 50 to 55% w/w carbon coal will have a density of approximately 1.5 g/cm^3, and this will decrease to, say, 1.3 g/cm^3 for an 85% carbon coal followed by an increase in density to about 1.8 g/cm^3 for a 97% carbon coal. On a comparative note, the density of graphite (2.25 g/cm^3) also falls into this trend.

Determinations of the density of various coal macerals have also been reported (Table 6.2), and although the variations are not great, the general order of density for macerals (having the same approximate carbon content) is

$$\text{exinite} < \text{vitrinite} < \text{micrinite} \tag{6.2}$$

However, it should also be noted that the density of a particular maceral does vary somewhat with the carbon content (Figure 6.2) (van Krevelen, 1957; Berkowitz, 1979).

TABLE 6.2 Alcohol and Helium Density of Macerals

Macerals	% C	D_{CH_3OH}	$d_{He(calc)}$
Exinites	85.49	1.201	1.187
	87.41	1.213	1.193
	89.10	1.288	1.267
	89.29	1.347	1.325
	83.5	1.345	1.304
	85.74	1.334	1.304
Vitrinites	88.36	1.317	1.295
	88.84	1.368	1.338
	86.77	1.463	1.435
	87.98	1.415	1.386
Micrinites	89.59	1.414	1.389
	89.78	1.413	1.385

Source: Adapted from Kroger and Badenecker (1957) and Braunstein et al. (1977).

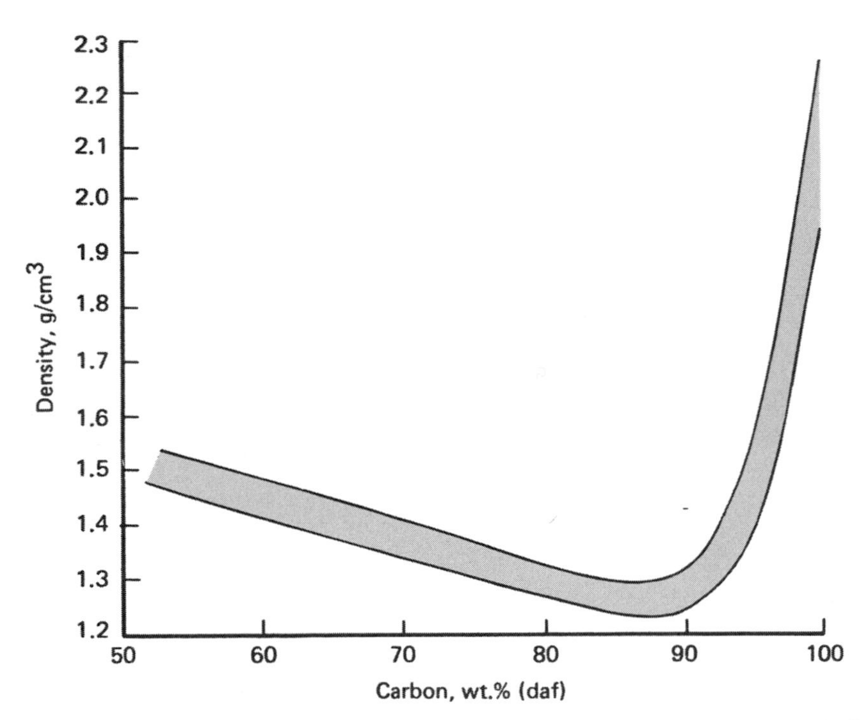

FIGURE 6.2 Variation of vitrinite density with carbon content. (Adapted from Berkowitz, 1979.)

TABLE 6.3 Variations of the Bulk Density of Coal for Different Size Fractions[a]

Particle Size of Coal Sample	Bulk Density (lb/ft^3)	Percent Voids
Mine run	55	37
Lump (plus 6 in.)	50	43
6×3 in.	48	45
3×2 in.	46	47
2-in. screenings	49	44
$2 \times 1\frac{1}{2}$ in.	45	48
$\frac{3}{4} \times \frac{7}{16}$ in.	42	52
$\frac{7}{16}$ in. $\times$ 0	47	46
$\frac{5}{16}$ in. $\times$ No. 10	45	48
$\frac{1}{16}$ in. $\times$ 48 M	42	52
No. 48 $\times$ 0	35	60

Source: Baughman (1978, p. 166).
[a] Specific gravity of coal: 1.4.

The *bulk density* is the mass of an assembly of coal particles in a container divided by the volume of the container (Table 6.3). It depends on true density, particle size and size distribution, particle shape, surface moisture, and degree of compaction. The parameter is often used in the design of handling, transportation, and storage systems for coal. The test method (ASTM D-291) concerns the compaction of crushed coal to determine either its compacted or uncompacted weight, for purposes such as charging coke ovens. In addition to the character of the coal itself, moisture content and size distribution of the coal are the two main factors that affect the cubic foot weight. A moisture determination and sieve analysis of the coal should be reported along with the cubic foot weight for proper interpretation of the cubic foot weight. During the period of collecting the gross sample, the increments of the sample should be stored in a waterproof container with a tightly fitting cover to prevent the loss of moisture.

The *in-place density* (*bank density*) of coal is the means by which coal in the seam can be expressed as tons per acre per foot of seam thickness and/or tons per square mile per foot of seam thickness (Table 6.4). The *in-place density* must be determined on water-saturated samples (Berkowitz, 1979) to accommodate the equilibrium moisture (Chapter 3) that exists under the in-place (or seam) conditions.

6.2 POROSITY AND SURFACE AREA

Porosity is the fraction (or percentage) of the volume of coal that is occupied by pores and can be calculated from the equilibrium moisture content (Chapter 3)

TABLE 6.4 In-Place Density of Coal

Rank	Weight of In-Place Coal: Tons per Acre per Foot of Thickness	Weight of In-Place Coal: Tons per Square Mile per Foot of Thickness
Anthracite	2310	1479×10^6
Semianthracite	2039	1305×10^6
Bituminous	1903	1218×10^6
Subbituminous	1767	1131×10^6
Lignite	1631	1044×10^6

Source: Adapted from Baughman (1978, p. 166).

(Berkowitz, 1979). Since coal is a porous material, porosity can have a large influence on coal behavior during mining, preparation, and utilization operations.

Although porosity dictates the rate at which methane can diffuse out of the coal (in the seam) and there may also be some influence during preparation operations in terms of mineral matter removal, the major influence of the porous nature of coal is seen during the utilization of coal. For example, during conversion, chemical reactions occur between gas (and/or liquid) products and surface features, much of which exists within the pore systems.

The calculation of porosity is derived from the determination of the true specific gravity (ASTM D-167) and is derived from the relationship

$$\text{porosity} = 100 - 100\left(\frac{\text{apparent specific gravity}}{\text{true specific gravity}}\right) \tag{6.3}$$

Another method of determining porosity involves measuring the density of coal by helium displacement and by mercury displacement (see Section 6.1). Thus, the porosity of coal is calculated from the relationship

$$P = 100 \times \text{density}_{\text{Hg}}\left(\frac{1}{\text{density}_{\text{Hg}}} - \frac{1}{\text{density}_{\text{He}}}\right) \tag{6.4}$$

where P is the porosity, $\text{density}_{\text{Hg}}$ the mercury density, and $\text{density}_{\text{He}}$ the helium density.

By determining the apparent density of coal in fluids of different but known dimensions, it is possible to calculate the pore size (pore volume) distribution. The open pore volume (V), the pore volume accessible to a particular fluid, can be calculated from the relationship

$$V = \frac{1}{\text{density}_{\text{Hg}}} - \frac{1}{\text{density}_{\text{a}}} \tag{6.5}$$

where $density_{Hg}$ is the mercury density and $density_{a}$ is the apparent density in the fluid under consideration.

Pore volume can be calculated from the relationship

$$V_p = \frac{1}{density_{Hg}} - \frac{1}{density_{He}} \tag{6.6}$$

where $density_{Hg}$ is the mercury density and $density_{He}$ is the helium density.

The pore systems of coal have generally been considered to consist of micropores having sizes up to approximately 100 Å and macropores having sizes greater than 300 Å (Gan et al., 1972; Mahajan and Walker, 1978). Other work (Kalliat et al., 1981), involving a small-angle x-ray investigation of porosity in coals, has thrown some doubt on this hypothesis by bringing forward the suggestion that the data are not consistent with the suggestion that many pores have dimensions some hundreds of angstrom units in diameter but have restricted access due to small openings which exclude nitrogen (and other species) at low temperatures. Rather, the interpretation that is favored is that the high values of surface area obtained by adsorption studies are the result of a large number of pores whose minimum pore dimensions are not greater than about 30 Å.

There are also indications that the adsorption of small molecules on coal, such as methanol, occurs by a site-specific mechanism (Ramesh et al., 1992). In such cases it appears that the adsorption occurs first at high-energy sites, but with increasing adsorption the (methanol) adsorbate continues to bind to the surface rather than to other (polar) methanol molecules, and there is evidence for both physical and chemical adsorption. In addition, at coverage below a monolayer, there appears to be an activation barrier to the adsorption process. Whether or not such findings have consequences for surface area and pore distribution studies remains to be seen. But there is the very interesting phenomenon of the activation barrier, which may also have consequences for the interpretation of surface effects during coal combustion (Speight, 1994). As an aside, adsorption studies of small molecules on coal has been claimed to confirm the copolymeric structure of coal (Milewska-Duda, 1991).

As already noted with respect to coal density (Figure 6.1), the porosity of coal decreases with carbon content (Figure 6.3) (King and Wilkins, 1944; Berkowitz, 1979) and has a minimum at approximately the 89% w/w carbon coals followed by a marked increase in porosity. The nature of the porosity also appears to vary with carbon content (rank); for example, the macropores are usually predominant in the lower carbon (rank) coals whereas higher carbon (rank) coals contain predominantly micropores. Thus, pore volume decreases with carbon content (Figure 6.4) and, in addition, the surface area of coal varies over the range 10 to 200 m^2/g and also tends to decrease with the carbon content of the coal.

Immersing the coal in mercury and increasing the pressure progressively can determine the size distribution of the pores within a coal. Surface tension effects prevent the mercury from entering pores having a diameter smaller than a given

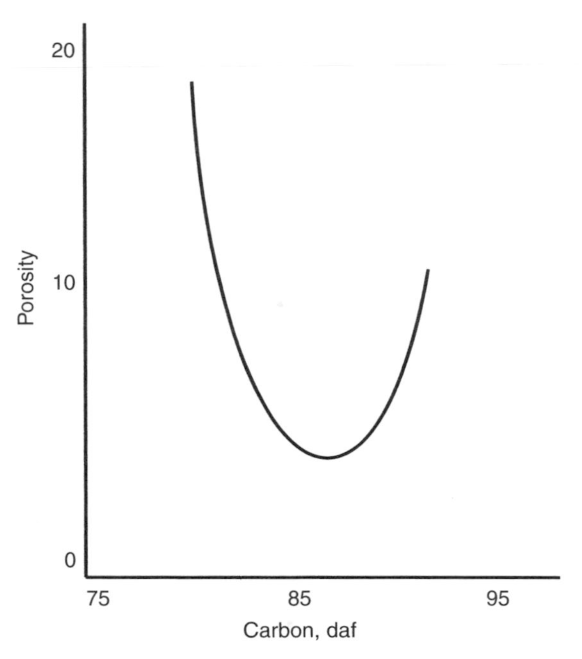

FIGURE 6.3 Variation of porosity with carbon content. (Adapted from Berkowitz, 1979.)

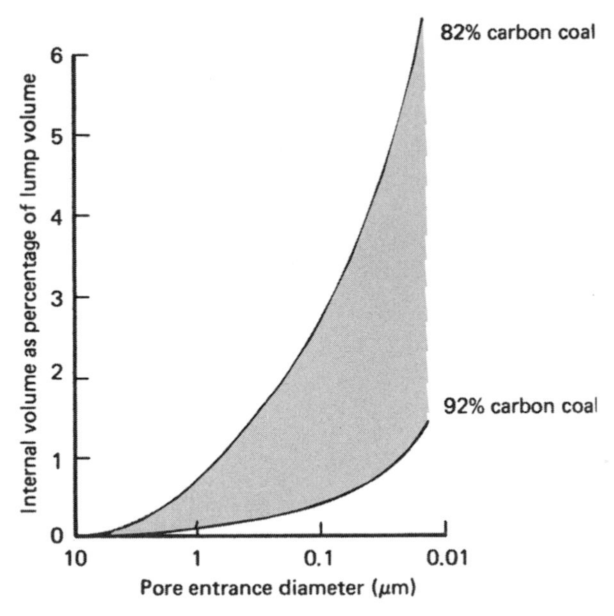

FIGURE 6.4 Variation of pore distribution with carbon content.

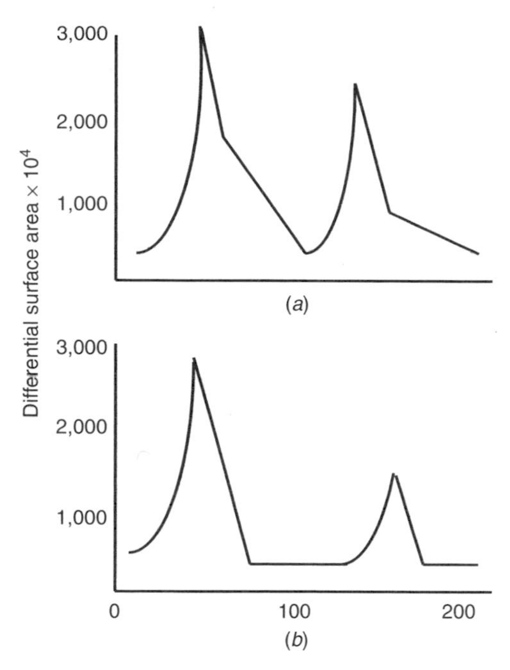

FIGURE 6.5 Variation of surface area with pore size. (Adapted from Berkowitz, 1979.)

value d for any particular pressure p such that

$$p = \frac{4\sigma \cos \theta}{d} \tag{6.7}$$

where σ is the surface tension and θ the angle of contact.

By recording the amount of mercury entering the coal for small increments of pressure, it is possible to build up a picture of the variation of surface area with pore size (Figure 6.5) (van Krevelen, 1957; Berkowitz, 1979). However, the total pore volume accounted for by this method is substantially less than that derived from the helium density, thereby giving rise to the concept that coal contains two pore systems: (1) a macropore system accessible to mercury under pressure, and (2) a micropore system that is inaccessible to mercury but accessible to helium. By using liquids of various molecular sizes, it is possible to investigate the distribution of micropore sizes. However, the precise role or function of these micropores as part of the structural model of coal is not fully understood, although it has been suggested that coal may behave in some respects like a molecular sieve.

6.3 REFLECTANCE

Coal, a black solid that *appears* to be impervious to light, does exhibit optical properties. However, to exhibit these properties, some preparation or conditioning of the coal is also of prime importance. Thus, coal may be examined in visible light by either transmission or reflectance. The former is a measure of light absorbance at various wavelengths and may be determined for thin sections of coal or finely divided coal pressed into a potassium bromide disk or solutions of coal extracts in solvents such as pyridine or films of coal that have been deposited by evaporation of a dispersing liquid.

Coal reflectance (ASTM D-2798) is very useful because it indicates several important properties of coal, including determination of the maceral composition of coal, which, in turn, is helpful for the prediction of behavior in processing (Davis, 1978; Davis et al., 1991). Coal reflectance is determined by the relative degree to which a beam of polarized light is reflected from a polished coal surface that has been prepared according to a standard procedure (ASTM D-2797). Samples prepared by this practice are used for microscopical determination of the reflectance of the organic components in a polished specimen of coal (ASTM D-2798) as well as the volume percent of physical components of coal (ASTM D-2799).

The coal is crushed to pass a number 20 (850-μm) screen (with minimal fines production) and the particles are formed into a briquette held together with a binder. One side of the briquette is polished using successively finer abrasives until a smooth surface that is scratch-free, smear-free, and char-free is obtained. A metallurgical, or opaque-ore, microscope is employed to determine the reflectance with vertical illumination using polarized light.

The standard test method for the measurement of vitrinite reflectance (ASTM D-2798) covers the microscopical determination of both the mean maximum reflectance and the mean random reflectance measured in oil of polished surfaces of vitrinite and other macerals present in coals ranging in rank from lignite to anthracite. This test method can be used to determine the reflectance of other macerals. The reflectance of the maceral vitrinite or other macerals is determined in this test method by illuminating a polished surface of a section of coal in immersion oil using a microscopic system that measures photometrically the amount of light reflected from the surface. The reflected light is recorded in percent reflectance after calibration of photometric equipment by measuring the reflected light from standards of reflectance as calculated from their refractive indexes. The mean maximum reflectance of the vitrinite component in coal as determined by this test method is often used as an indicator of rank (ASTM D-388), independent of petrographic composition, and in the characterization of coal as feedstock for carbonization, gasification, liquefaction, and combustion processes. If mean maximum reflectance is used as a rank indicator, the types of vitrinite measured should be specified.

Prior to measurement of reflectance, the sample face is covered with cedar oil or commercial immersion oil and then multiple readings are taken of the maximum reflectance of the coal component (e.g., vitrinite) of interest. These values are compared with readings of high-index glass standards (of known reflectance) that are available with reflectance values typically ranging from 0.302 to 1.815%.

Thus, although coal always appears as a black mass, thin layers and polished faces exhibit a variety of colors. For example, in incident light, fusinite and micrinite are white, whereas exinite is a translucent yellow color; on the other hand, exinite is orange in transmitted light. Obviously, these color differences are employed for differentiating maceral types. In addition, the reflectance of coal varies with carbon content (Figure 6.6) and the reflectance data for air are considerably higher than those obtained using an oil medium (Table 6.5).

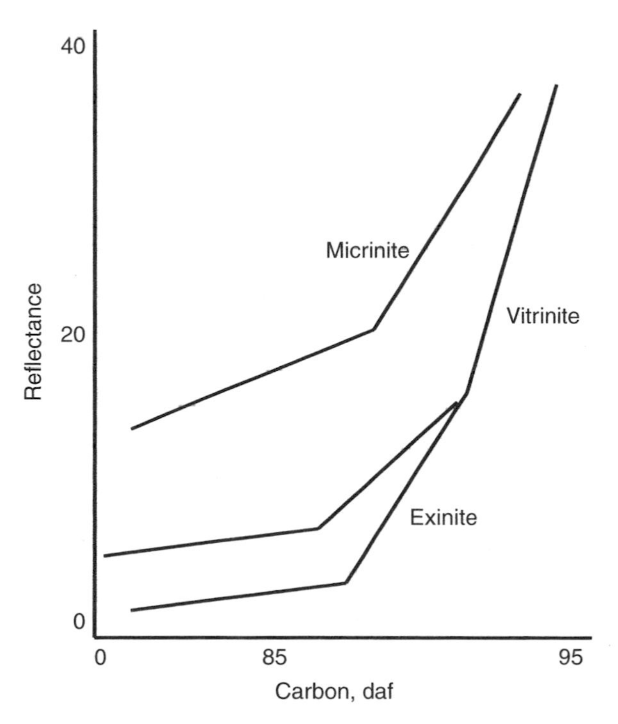

FIGURE 6.6 Variation of reflectance with carbon content. (Adapted from Dorman et al., 1957.)

TABLE 6.5 Reflectance Data Obtained in Air and in Oil

	Maximum Reflectance, Approximate (%)	
Carbon Content of Coal (%)	Air	Oil
60	0.4	6
90	1.0	8
96	6.5	17

The volume percent of physical components of coal is used as an aid in the characterization of coal for use in carbonization, gasification, and combustion processes. To determine the volume percent of the physical components of coal (ASTM D-2799), the components in a representative crushed coal sample (ASTM D-2797) are identified under a microscope according to their reflectance, other optical properties, and morphology. The proportion of each component in the sample is determined by observation of a statistically adequate number of points and summing those representative of each component. Only area proportions of components are determined on a surface section of a sample. However, the area and volume proportions are the same when the components are distributed randomly throughout the sample.

6.4 REFRACTIVE INDEX

The refractive index of coal can be determined by comparing the reflectance in air with that in cedar oil. A standard test method (ASTM D-2798) covers the microscopic determination of both the mean maximum reflectance and the mean random reflectance measured in oil of polished surfaces of vitrinite and other macerals in coal ranging in rank from lignite to anthracite. This test method can be used to determine the reflectance of other macerals. For vitrinite (various coals), the refractive index usually falls within the range 1.68 (58% carbon coal) to 2.02 (96% carbon coal).

6.5 CONDUCTIVITY

The electrical conductivity of coal is generally discussed in terms of specific resistance, p (units of p are ohm-centimeters), and is the resistance of a block of coal 1 cm long and a 1 cm^2 in cross section. Substances having a specific resistance greater than approximately $1 \times 10^{15}\ \Omega \cdot cm$ are classified as insulators, and those with a specific resistance below $1\ \Omega \cdot cm$ are *conductors*; materials between these limits are *semiconductors*.

Electrical conductivity depends on several factors, such as temperature, pressure, and moisture content of the coal. The electrical conductivity of coal is quite pronounced at high temperatures [especially above 600°C (1112°F)], where coal structure begins to break down. Moisture affects electrical conductivity to a marked extent, resulting in a greatly increased conductivity. To prevent any anomalies from the conductance due to water, the coal is usually maintained in a dry, oxygen-free atmosphere, and to minimize the problems that can arise, particularly because of the presence of water, initial measurements are usually taken at approximately 200°C (392°F) and then continued to lower temperatures.

Coal is a semiconductor (van Krevelen, 1961; Speight, 1994, and references cited therein). Anthracite is a semiconductor with specific resistance ranging from 1 to 104 $\Omega \cdot cm$, whereas the range is 105 to 1012 $\Omega \cdot cm$ for bituminous coal.

Subbituminous coal also behaves as a semiconductor. The highest resistances are observed with coals having 80 to 92% carbon; they can be considered virtual insulators. Hence, electrical conductivity, a measure of electricity transportation, is generally handled in terms of electrical resistivity for coal. To eliminate quantities not characteristic for the material, the specific electrical resistance (resistivity) and the specific conductance are utilized. The former is the resistance that a cube with unit dimensions offers to current flow and is expressed in ohm-unit length, and the latter is the reciprocal of the former. Taking the expected order of magnitude of resistivity or conductivity into account, measurements can be made with a fluxmeter, an ammeter-voltmeter system, a Wheatstone bridge, and electrometers (ASTM D-257) for the determination of direct-current insulation resistance and are strongly recommended for transient resistivity measurements. The test method covers direct-current procedures for the determination of dc insulation resistance, volume resistance, volume resistivity, surface resistance, and surface resistivity of electrical insulating materials, or the corresponding conductance and conductivities. The method is not suitable for use in measuring the electrical resistivity or conductivity of moderately conductive materials, for which an alternative method (ASTM D-4496) is recommended.

ASTM D-4496 covers determination of the measurement of electrical resistance or conductance of materials that are generally categorized as moderately conductive and are neither good electrical insulators nor good conductors. This method applies to all materials that exhibit volume resistivity in the range 1 to 10 $\Omega \cdot$ cm or surface resistivity in the range of 10 to 10 Ω (per square). The method is designed for measurements at standard conditions of 23°C and 50% relative humidity, but its principles of operation can be applied to specimens measured at lower or higher temperatures and relative humidity.

All electrical property values are strongly dependent on water content; for water, the dielectric constant is approximately 81 and resistivity is about 106 $\Omega \cdot$ cm. The dielectric constant has been used as a measure of moisture in coal (Speight, 1994, and references cited therein). However, it should be noted that the effect is not considered to be additive due to the different electrical properties of physically and chemically bound water. With an increase in moisture content, electrical conductivity and dielectric constant increase, whereas resistivity and dielectric strength decrease. Hence, except for special purposes (e.g., dielectric strength measurements of underground coal blocks), electrical measurements require the meticulous drying of coal prior to experiments.

The properties outlined in this chapter must always be borne in mind when consideration is being given to the suitability of coal for a particular use. It must also be borne in mind that coal which at first appears unsuitable for use by a consumer might become eminently suitable by a simple or convenient pretreating step, almost analogous to the *conditioning* of asphalt by air blowing (Speight, 1999).

The conductivity of coal is explained in part by the partial mobility of electrons in the coal structure lattice which occurs because of unpaired electrons or "free radicals." Mineral matter in coal may have some influence on electrical conductivity. The conductivities of coal macerals show distinct differences; fusains conduct electricity much better than do clarain, durain, and vitrain.

6.6 DIELECTRIC CONSTANT

The dielectric constant (specific inductive capacity) is a measure of electrostatic polarizability and of the amount of electricity that can be stored in coal. The dielectric constant is more useful than electrical conductivity in characterizing coal and is a measure of the electrostatic polarizability of the dielectric coal. The dielectric constant of coal is believed to be related to the polarizability of the π-electrons in the clusters of aromatic rings in the coal chemical structure.

Experimental methods are applicable for a wide range of frequencies. High-frequency measurements employ commercially available dielectric constant meters, Q-meters, and so on; the impedance bridge method is widely employed at low frequencies. The levels of the frequencies applied experimentally are very important for data interpretation and comparison.

The dielectric constant of coal is strongly dependent on coal rank (van Krevelen, 1961; Speight, 1994, and reverences cited therein). For dry coals the minimum dielectric constant value is 3.5 and is observed at about 88% w/w carbon content in the bituminous coal range. The dielectric constant increases sharply and approaches 5.0 for both anthracite (92% carbon) and lignite (70% carbon). The Maxwell relation which equates the dielectric constant to the square of the refractive index for a polar insulators generally shows a large disparity even for strongly dried coal.

Like conductivity, dielectric constant is strongly dependent on water content. Indeed, the dielectric constant can even be used as a measure of moisture in coal (Chapter 3). Meaningful dielectric constant measurements of coal require drying to a constant dielectric constant, and several forms of coal are used for dielectric constant measurements. These include precisely shaped blocks of coal, mulls of coal in solvents of low dielectric constant, or blocks of powdered coal in a paraffin matrix.

The dielectric constant varies with coal rank (Chatterjee and Misra, 1989). The theorem that the dielectric constant is equal to the square of the refractive index (which is valid for nonconducting, nonpolar substances) holds only for coal at the minimum dielectric constant. The decreasing value of dielectric constant with rank may be due to the loss of polar functional groups (such as hydroxyl or carboxylic acid functions), but the role of the presence of polarizable electrons (associated with condensed aromatic systems) is not fully known. It also appears that the presence of intrinsic water in coal has a strong influence on the dielectric properties (Chatterjee and Misra, 1989).

6.7 DIELECTRIC STRENGTH

Dielectric strength indicates the voltage gradient at which dielectric failure occurs is generally measured (ASTM D-149) at commercial power frequencies, and is expressed in volts per sample thickness. This test method covers procedures for the determination of dielectric strength of solid insulating materials at commercial power frequencies, under specified conditions. Unless otherwise specified, the tests are conducted at 60 Hz. However, this test method may be used at any frequency from 25 to 800 Hz but at frequencies above 800 Hz dielectric heating may occur. The test method is not intended for determining intrinsic dielectric strength, direct-voltage dielectric strength, or thermal failure under electrical stress (see ASTM D-3151). The test method is most commonly used to determine the dielectric breakdown voltage through the thickness of a test specimen (puncture). It may also be used to determine dielectric breakdown voltage along the interface between a solid specimen and a gaseous or liquid surrounding medium (flashover).

Experimental data strongly indicates that anthracite and bituminous coal are electrically anisotropic. Higher resistivity/lower conductivity is observed for specimens oriented perpendicular to the bedding plane relative to those with parallel orientation.

6.8 SPECIFIC RESISTANCE (RESISTIVITY)

Specific resistance (resistivity) is the electrical resistance of a body of unit cross section and of unit length, expressed in ohm-centimeters:

$$p = \frac{RA}{L} \tag{6.8}$$

where p is the specific resistance, R the resistance of the substance, A the cross-sectional area, and L the length. The specific resistance of coal (Table 6.6) may vary from thousands of ohm-centimeters to millions of ohm-centimeters, depending on the direction of measurement.

When wet coal is exposed to higher temperatures (0 to 200°C, 32 to 392°F), an increase in electrical resistivity (with a concurrent decrease of dielectric constant) is observed. This is due to moisture loss. After moisture removal, a temperature increase results in lower resistivity (and higher dielectric constant). The dependency of conductive properties on temperature is mainly exponential, as in any semiconductor. At lower temperatures, the effect of temperature on electrical properties is reversible. The onset of irreversible effects is rank dependent and starts at 200 to 400°C (392 to 752°F) for bituminous coal and at 500 to 700°C (932 to 1292°F) for anthracite.

The effect of volatile matter content on electrical properties is not fully understood. Some data indicate that there is no particular correlation, whereas other data indicate a slight decrease of resistivity with decreasing volatile matter content. Both the type and quantity of volatile matter may affect electrical properties. However, it should be noted that even samples with equal volatile matter may show different properties due to different origins. Both mineral matter

TABLE 6.6 Specific Resistance of Coal

Material	Specific Resistance ($\Omega \cdot cm$)
Graphite	$(0.8–1.0) \times 10^{-3}$
Graphite	4×10^{-3}
Anthracite[a]	
Parallel to bedding	$(7–90) \times 10^{3}$
Perpendicular to bedding	$(17–34) \times 10^{3}$
Bituminous[a]	
Parallel to bedding	$(0.004–360) \times 10^{8}$
Perpendicular to bedding	$(3.1–530) \times 10^{9}$
Brown coal	
20–25% H_2O	10^{4}
Dry	$10^{10}–10^{13}$
Copper	1.7×10^{-6}
Water (distilled)	$8.5–25 \times 10^{6}$

Source: Adapted from Baughman (1978, p. 170).
[a]Dry.

and ash content affect electrical properties, the influence of the former being more pronounced).

REFERENCES

Agrawal, P. L. 1959. *Proceedings of the Symposium on the Nature of Coal*. Central Fuel Research Institute, Jealgora, India, p. 121.

ASTM, 2004. *Annual Book of ASTM Standards*, Vol. 05.06. American Society for Testing and Materials, West Conshohocken, PA. Specifically:

ASTM D-149. Standard Test Method for Dielectric Breakdown Voltage and Dielectric Strength of Solid Electrical Insulating Materials at Commercial Power Frequencies.

ASTM D-167. Standard Test Method for Apparent and True Specific Gravity and Porosity of Lump Coke.

ASTM D-346. Standard Practice for Collection and Preparation of Coke Samples for Laboratory Analysis.

ASTM D-257. Standard Test Methods for DC Resistance or Conductance of Insulating Materials.

ASTM D-291. Standard Test Method for Cubic Foot Weight of Crushed Bituminous Coal.

ASTM D-388. Standard Classification of Coals by Rank.

ASTM D-2797. Standard Practice for Preparing Coal Samples for Microscopical Analysis by Reflected Light.

ASTM D-2798. Standard Test Method for Microscopical Determination of the Reflectance of Vitrinite in a Polished Specimen of Coal.

ASTM D-2799. Standard Test Method for Microscopical Determination of Volume Percent of Physical Components of Coal.

ASTM D-3151. Standard Test Method for Thermal Failure of Solid Electrical Insulating Materials Under Electric Stress.

ASTM D-4496. Standard Test Method for D-C Resistance or Conductance of Moderately Conductive Materials.

Baughman, G. L. 1978. *Synthetic Fuels Data Handbook.* Cameron Engineers, Denver, CO.

Berkowitz, N. 1979. *An Introduction to Coal Technology*. Academic Press, San Diego, CA.

Braunstein, H. M., Copenhauer, E. D., and Pfuderer, H. A. 1977. *Environmental Health and Control Aspects of Coal Conversion: An Information Review*. Report ORNL-EIS-94. Oak Ridge National Laboratory, Oak Ridge, TN.

Chatterjee, I., and Misra, M. 1989. *Proc. Mater. Res. Symp. Mater. Res. Soc.*, 189:195.

Davis, A. 1978. In *Analytical Methods for Coal and Coal Products*, Vol. 1, C. Karr, Jr. (Editor). Academic Press, San Diego, CA, Chap. 2.

Davis, A., Mitchell, G. D., Derbyshire, F. J., Rathbone, R. F., and Lin, R. 1991. *Fuel*, 70:352.

Dorman, H. N. M., Huntjens, F. J., and van Krevelen, D. W. 1957. *Fuel*, 36:321.

Ergun, S. 1951. *Anal. Chem.*, 23:151.

Ettinger, I. L., and Zhupakhina, E. S. 1960. *Fuel*, 39:387.

Evans, D. G., and Allardice, D. J. 1978. In *Analytical Methods for Coal and Coal Products*, Vol. 1, C. Karr, Jr. (Editor). Academic Press, San Diego, CA, Chap. 3.

Gan, H., Nandi, S. P., and Walker, P. L., Jr. 1972. *Fuel*, 51:272.

Kalliat, M., Kwak, C. Y., and Schmidt, P. W. 1981. In *New Approaches in Coal Chemistry*, B. D. Blaustein, B. C. Bockrath, and S. Friedman (Editors). American Chemical Society, Washington, DC, Chap. 1.

King, J. G., and Wilkins, E. T. 1944. *Proceedings of the Conference on the Ultrafine Structure of Coal and Cokes*. British Coal Utilization Research Association, London, p. 45.

Kroger, C., and Badenecker, J. 1957. *Brennst. Chem.*, 38:82.

Labuschagne, B. C. J. 1987. *Proceedings of the 4th Annual Pittsburgh Coal Conference*. University of Pittsburgh, Pittsburgh, PA, p. 417.

Labuschagne, B. C. J., Wheelock, T. D., Guo, R. K., David, H. T., and Markuszewski, R. 1988. *Proceedings of the 5th Annual International Pittsburgh Coal Conference*. University of Pittsburgh, Pittsburgh, PA, p. 417.

Lyman, W. J., Reehl, W. F., and Rosenblatt, D. H. 1990. *Handbook of Chemical Property Estimation Methods: Environmental Behavior of Organic Compounds*. McGraw-Hill, New York.

Mahajan, O. P., and Walker, P. L., Jr. 1978. In *Analytical Methods for Coal and Coal Products*, Vol. 1, C. Karr, Jr. (Editor). Academic Press, San Diego, CA, Chap. 4.

Milewska-Duda, J. 1991. *Arch. Mineral. Sci.*, 36:369.

Ramesh, R., Francois, M., Somasundaran, P., and Cases, J. M. 1992. *Energy Fuels*, 6:239.

Speight, J. G. 1978. In *Analytical Methods for Coal and Coal Products*, Vol. 2, C. Karr, Jr. (Editor). Academic Press, San Diego, CA, Chap. 22.

Speight, J. G. 1994. *The Chemistry and Technology of Coal*, 2nd ed. Marcel Dekker, New York.

Speight, J. G. 1999. *The Chemistry and Technology of Petroleum*, 3rd ed. Marcel Dekker, New York.

van Krevelen, D. W. 1957. *Coal: Aspects of Coal Constitution*. Elsevier, Amsterdam.

Walker, P. L., Jr. 1981. *Philos. Trans. R. Soc. (London)*, 300A:65.

7 Thermal Properties

The thermal properties of coal are important in determining the applicability of coal to a variety of conversion processes. For example, the *heat content* (also called the *heating value* or *calorific value*) is often considered to be the most important thermal property. However, there are other thermal properties that are of importance insofar as they are required for the design of equipment that is to be employed for the utilization (conversion, thermal treatment) of coal in processes such as combustion, carbonization, gasification, and liquefaction. Plastic and agglutinating properties as well as phenomena such as the agglomerating index give indications of how coal will behave in a reactor during a variety of thermal processes (Chan et al., 1991).

7.1 CALORIFIC VALUE

The calorific value is the heat produced by the combustion of a unit quantity of coal in a bomb calorimeter with oxygen and under a specified set of conditions (ASTM D-121; ASTM D-2015; ASTM D-3286; ISO 1928). For the analysis of coal, the calorific value is determined in a bomb calorimeter either by a static (isothermal) method or by an adiabatic method, with a correction made if net calorific value is of interest. The unit is calories per gram, which may be converted to the alternate units (1.0 kcal/kg = 1.8 Btu/lb = 4.187 kJ/kg).

The calorific value is a direct indication of the heat content (energy value) of the coal and represents the combined heats of combustion of the carbon, hydrogen, nitrogen, and sulfur in the organic matter and of the sulfur in pyrite and is the gross calorific value with a correction applied if the net calorific value is of interest.

The calorific value is usually expressed as the *gross calorific value* (GCV) or the *higher heating value* (HHV) and the *net calorific value* (NCV) or *lower calorific value* (LHV). The difference between the gross calorific value and the net calorific value is the latent heat of condensation of the water vapor produced during the combustion process. The gross calorific value assumes that all of the vapor produced during the combustion process is fully condensed. The net calorific value assumes that the water is removed with the combustion products without being fully condensed. To equalize all effects, the calorific value of coal

Handbook of Coal Analysis, by James G. Speight
ISBN 0-471-52273-2

should be compared based on the net calorific value basis. The calorific value of coal varies considerably, depending on the ash, moisture content, and the type of coal, whereas the calorific value of fuel oils is much more consistent.

The calorific value of coal is an important property. For example, the gross calorific value can be used to compute the total calorific content of the quantity of coal or coke represented by the sample for payment purposes. It can also be used to compute the calorific value versus sulfur content to determine whether the coal meets regulatory requirements for industrial fuels. The gross calorific value can be used to evaluate the effectiveness of beneficiation processes. Finally, the gross calorific value can be required to classify coal (ASTM D-388).

The energy content of the coal can be expressed as the *useful heating value* (UHV), which is an expression derived from the ash and moisture contents for noncaking coal through the formula

$$\text{UHV (kcal/kg)} = 8900 - 138 \times [\text{ash content (wt \%)} + \text{moisture content (wt \%)}] \qquad (7.1)$$

The calorific value is neither part of the proximate analysis nor part of the ultimate analysis; it is, in fact, one of the many physical properties of coal and as such, is often found in the various sections that deal with the physical properties of coal. In the present context, the importance of the calorific value as one of the means by which coal can be evaluated dictates that it be included in this particular section as well as in the section describing the general thermal properties of coal.

Calorimeters used for determination of the calorific value of coal can be classified into three general types: (1) calorimeters using solid oxidizing agents in either open or closed systems, (2) calorimeters using gaseous oxygen at approximately atmospheric pressure in an open system, and (3) calorimeters using gaseous oxygen under pressure in a closed system. All three types of calorimeters are commonly used, but only those belonging to the third class can be relied on for a reasonable degree of accuracy.

The bomb calorimeter provides the most suitable and accurate apparatus for determination of the calorific values of solid and liquid fuels. Since the combustion takes place in a closed system, heat transfer from the calorimeter to the water is complete, and since the reaction is one between the fuel and gaseous oxygen, no corrections are necessary for the heat absorbed during the reduction of the oxidizing agent. In addition, the losses due to radiation can be reduced to comparatively small quantities, and more important, can be determined with a considerable degree of accuracy. Corrections due to the heat evolved in the formation of nitric and sulfuric acids under the conditions existing in the bomb can be determined accurately.

When coal is burned on an open grate under normal air pressures, only minute traces of oxides of nitrogen are formed by the combination with atmospheric oxygen of nitrogen in the fuel or in the air, but under the conditions of local high temperature and pressure existing within the bomb, more oxides of nitrogen are formed, and these give rise ultimately to a solution containing nitric acid by dissolving in the water. This oxidation of nitrogen and solution of the products in water are exothermic reactions, so an allowance has to be made for the heat liberated by the formation of nitric acid.

In a similar way, when sulfur or pyrites burns in air, only about 5% of the sulfur is oxidized to sulfur trioxide, the remainder yielding only sulfur dioxide, but in the bomb practically all the sulfur is burned to the trioxide, which dissolves in the water to give sulfuric acid. A correction is therefore necessary for heat liberated by the combustion of sulfur dioxide to trioxide, by the solution of this trioxide in water to give sulfuric acid, and by the heat of dilution of this acid in water.

Applying a correction to the temperature rise observed makes an allowance for losses of heat by radiation from the calorimeter. This correction is expressed by a formula that appears to be complicated, but if the various temperature figures are set out on the laboratory sheet methodically, it is not only easy to apply, but the liability of introducing arithmetic errors is reduced.

The Beckmann thermometer used with the bomb calorimeter should be calibrated for the normal depth of immersion with which it is used. To cover the normal range of laboratory temperatures, this calibration should be obtained for three settings of the zero on the scale, convenient values being 10, 15, and 20°C. Such a series of calibrations allows automatically for emergent stem corrections and variations in the value of the degree on the thermometer scale with different quantities of mercury in the bulb, in addition to those arising from inherent variations in the diameter of the capillary bore.

Anthracite may be difficult to burn completely, but combustion is assisted by the addition of a small known amount of liquid paraffin of known calorific value. It is also advantageous to increase the pressure of oxygen to 30 atm.

If the coal ash has a low fusion point, the fused residue may enclose unburned material. In such cases the provision of a thin layer of coarsely powdered quartz or even a calcined asbestos paper disk placed on the bottom of a crucible prevents the formation of large globules of fused ash and therefore reduces the possibility of carbon remaining unburned. By using these modifications, the residue is more easily removed from the crucible (especially if it is made of silica) for examination for unburned carbon.

If the determination of calorific value is carried out with a bomb calorimeter, the figure reported is usually corrected for the heat of formation of sulfuric acid and its dilution by water, but if the calorific value has to be reported on a dry, mineral-matter-free basis, the air-dried figure must be adjusted further to allow for the heat of formation of ferric oxide from the pyrites originally in the coal. The heat developed in the combustion of pyrite (FeS_2) is approximately 3000 calories per gram of pyrite sulfur, so the correction to be applied is dependent on whether the calorific value is expressed as the heating power per gram or per pound.

The enthalpy, or heat content, of various coals has also been reported (Table 7.1) but has actually received somewhat less attention than the calorific

TABLE 7.1 Variation of Heat Content of Coal with Temperature

Temperature		Heat Content		
		As Tested		Ash-Free Basis
°C	°F	cal/g	Btu/lb	cal/g
		Lignite (Texas)		
32.7	90.9	11.8	21.2	13.5
69.3	156.7	20.2	36.4	22.5
95.3	203.5	25.4	45.7	27.7
34.4	273.9	39.2	70.6	42.5
		Subbituminous B (Wyoming)		
42.3	108.1	14.1	25.4	14.5
65.0	149.0	19.4	34.9	19.8
89.7	193.5	26.4	47.5	26.9
112.6	234.7	34.0	61.2	34.6

Source: Baughman (1978, p. 173).

value. There are also reports of the use of differential thermal analysis for the determination of the calorific value of coal (Munoz-Guillena et al., 1992).

7.1.1 Determination of Calorific Value

The common method of determining the gross calorific value of coal is with either an adiabatic calorimeter (ASTM D-2015; this test method was discontinued without replacement in 2000 but is still in use in many laboratories) or an isothermal bomb calorimeter (ASTM D-3286). In these methods, a weighed sample is burned in an oxygen bomb covered with water in a container surrounded by a jacket.

An adiabatic calorimeter is a calorimeter that has a jacket temperature adjusted to follow the calorimeter temperature so as to maintain zero thermal head, and the test method (ASTM D-2015, 1S0 1928) consists of burning the coal sample in the calorimeter, and the jacket temperature is adjusted during the burning so that it is essentially the same as the calorimeter water temperature. The calorific value is calculated from observations made before and after the combustion. In the isothermal method (ASTM D-3286; ISO 1928), the calorific value is determined by burning a weighed sample of coal in oxygen under controlled conditions, and the calorific value is computed from temperature observations made before, during, and after combustion with appropriate allowances made for the heat contributed by other processes. The value computed for the calorific value of coal is usually expressed in British thermal units per pound, kilocalories per kilogram, or in kilojoules per kilogram (1.8 Btu/lb = 1.0 kcal/kg = 4.187 kJ/kg).

In the isothermal calorimeter system (ASTM D-3286), the temperature rise of the calorimeter water is corrected for the heat lost to or gained from the surrounding jacket during the burning of the sample. In both systems, the corrected temperature rise times the energy equivalent of the calorimeter gives the total amount of heat produced during the burning of the sample. The *energy equivalent* (also called the *water equivalent* or *heat capacity*) of the calorimeter is determined by burning standard samples of benzoic acid.

After firing, the contents of the bomb are washed into a beaker and titrated with standard sodium carbonate solution to determine the amount of acid (nitric acid, HNO_3 and sulfuric acid, H_2SO_4) produced in the combustion. Corrections for the amount of acid, the amount of fuse wire used in firing, and the sulfur content of the sample are then made to the total heat produced in the calorimeter (energy equivalent times corrected temperature rise) to determine the gross calorific value of the solid fuel.

One source of error in this method (ASTM D-3286) is in temperature measurement. If a mercury-in-glass thermometer is used, it must be calibrated accurately and consistent readings must be made. Many calorimeters are equipped with digital thermometers with thermistor probes and microprocessors to control the firing and record the temperatures at prescribed intervals. This alleviates most of the human error in recording the temperature changes.

Igniting the coal sample in the oxygen bomb can be difficult. The sample may be blown out of the crucible by introducing the oxygen too quickly. Pressing the coal sample into a pellet may prevent the sample from blowing out. Coal with a high mineral content is hard to ignite, and mixing the sample with a measured amount of a standard aromatic acid and pelletizing may be helpful.

After firing, restoring the bomb pressure to atmospheric pressure too rapidly may result in the loss of oxides of sulfur and nitrogen. A correction must be made to the gross calorific value for the amounts of these acid-forming oxides produced in the bomb. Their loss results in a high calorific value. The pressure of the bomb must be restored very slowly to prevent this.

The equipment used must be checked periodically for any changes in the energy equivalent of the calorimeter, any corrosion or damage to the calorimeter bucket, any damage (however slight it may be) to the oxygen bomb, and any malfunction of the stirrers, electrical system, or other parts of the calorimeter. Any of these changes or malfunctions may change the energy equivalent of the calorimeter or introduce extra heat, which would lead to errors in the calorific value measured.

The experimental conditions require an initial oxygen pressure of 300 to 600 psi and a final temperature in the range 20 to 35°C (68 to 95°F) with the products in the form of ash, water, carbon dioxide, sulfur dioxide, and nitrogen. Thus, once the gross calorific value has been determined, the net calorific value (i.e., the net heat of combustion) is calculated from the gross calorific value [at 20°C (68°F)] by deducting 1030 Btu/lb (2.4×10^3 kJ/kg) to allow for the heat of vaporization of the water. The deduction is not actually equal to the heat of vaporization of water [1055 Btu/lb (2.45×10^3 kJ/kg)] because the calculation is to reduce the data from a gross value at constant volume to a net value at constant pressure. Thus, the differences between the gross calorific value (GCV) and the net calorific value (NCV) are given by

$$\text{NCV (Btu/lb)} = \text{GCV} - \frac{1030 \times \%\ \text{total hydrogen} \times 9}{100} \qquad (7.2)$$

In either form of measurement (ASTM D-2015; ASTM D-3286), the calorific value recorded is the gross calorific value. The net calorific value is calculated from the gross calorific value [at 20°C (68°F)] by making a suitable subtraction (= 1030 Btu/lb = 572 cal/g = 2.395 MJ/g) to allow for the water originally present as moisture as well as the moisture formed from the coal during the combustion. The deduction, however, is not equal to the latent heat of vaporization of water [1055 Btu/lb (2.4 MJ/g) at 20°C (68°F)] because the calculation is made to reduce from the gross value at constant volume to a net value at constant pressure for which the appropriate factor under these conditions is 1030 Btu/lb (2.395 MJ/g).

A more recent test method (ASTM D-5865) pertains to determination of the calorific value of coal by either an isoperibol (a calorimeter that has a jacket of uniform and constant temperature) or adiabatic bomb calorimeter.

For the adiabatic calorimeter, the jacket temperature must be adjusted to match that of the calorimeter vessel temperature during the period of the rise. The two temperatures must be maintained as close to equal as possible during the period of rapid rise. For the isoperibol calorimeter, the temperature rise may require a radiation correction. In either case, an individual test should be rejected if there is evidence of incomplete combustion. Furthermore, although it is required to check the heat capacity only once a month, this may be inadequate. A more frequent check of heat capacity values is recommended for laboratories making a large number of tests on a daily basis. The frequency of the heat capacity check should be determined to minimize the number of tests that would be affected by an undetected shift in the heat capacity values.

For all measurements of calorific value, caution is necessary during the sample preparation since oxidation of coal after sampling can result in a reduction of calorific value. In particular, lignite and subbituminous rank coal samples may experience greater oxidation effects than those of samples of higher-rank coals. Unnecessary exposure of the samples to the air for the time of sampling or delay in analysis should be avoided. Because of the nature of the test method, the calorimeter parts should be inspected carefully after each use and any calorimeter that has been dropped should not be used without a compete inspection of every part. Finally, the repeatability intervals of both methods of measurement are close, but there may be a human factor that can be overcome using a calorimeter with a programmer and digital thermometer.

If a coal does not have a measured calorific value, it is possible to make a close estimation of the calorific value (CV) by means of various formulas, the most popular of which are (Selvig, 1945):

1. The Dulong formula:

$$CV = 144.4(\%C) + 610.2(\%H) - 65.9(\%O) - 0.39(\%O)^2 \qquad (7.3)$$

2. the Dulong–Berthelot formula:

$$CV = 81{,}370 + 345\left[\frac{\%H - (\%O + \%N - 1)}{8}\right] + 22.2(\%S) \qquad (7.4)$$

where %C, %H, %N, %O, and %S are the respective carbon, hydrogen, nitrogen, oxygen, and organic sulfur contents of the coal (all of which are calculated to a dry, ash-free basis). In both cases, the values calculated are in close agreement with the experimental calorific values.

In some instances, an adiabatic bomb calorimeter may not be available or the sample may be too small for accurate use. To combat such problems, there is evidence that differential thermal analysis (DTA) is applicable to the determination of the calorific value of coal. Data obtained by use of the DTA method are in good agreement with those data obtained by use of the bomb calorimeter (Munoz-Guillena et al., 1992).

The heating value of coal is a function of rank and the standard classification of coals by rank according to their degree of coalification in ascending order from lignite to anthracite uses the heating value component (ASTM D-388). This classification takes into account fixed carbon and calorific value on a mineral-matter-free basis and agglomerating character determined by examination of the residue from the volatile determination. In high-volatile, medium-volatile, and low-volatile bituminous coals, the moisture and ash-free Btu (MAF Btu) commonly reported by commercial laboratories tends to increase with decreasing volatile matter from approximately 14,700 Btu/lb to approximately 15,600 Btu/lb. In the lower-rank coals the moisture, ash-free Btu decreases with decreasing rank, approaching roughly that of wood, in the range 6000 to 8000 Btu/lb for lignite.

7.1.2 Data Handling and Interpretation

The calorific value is normally the basic item specified in contracts for coal to be used in steam plants. It is the most important value determined for coal that is to be used for heating purposes. In coal contracts, the calorific value is usually specified on the as-received basis. Any error in the moisture value is reflected in the as-received calorific value.

The laboratory-determined calorific value is called the *gross calorific value*. It is the heat produced by combustion of a unit quantity of coal at constant volume in an oxygen bomb calorimeter under specified conditions such that the end products of the combustion are in the form of ash, gaseous carbon dioxide, sulfur dioxide, nitrogen, and liquid water. Burning coal as a fuel does not produce as much heat per unit quantity. Corrections are made to the gross calorific values for this difference between the laboratory and coal-burning facility. The corrected value is referred to as the *net calorific value*. This is defined as the heat produced by combustion of a unit quantity of coal at constant atmospheric pressure under conditions such that all water in the products remains in the form of vapor. The net calorific value is lower than the gross calorific value.

7.2 HEAT CAPACITY

The heat capacity of a material is the heat required to raise the temperature of 1 unit weight of a substance 1 degree, and the ratio of the heat capacity of one substance to the heat capacity of water at 15°C (60°F) in the specific heat.

The heat capacity of coal can be measured by standard calorimetric methods that have been developed for other materials (e.g., ASTM C-351). The units for heat capacity are Btu per pound per degree Fahrenheit (Btu/lb-°F) or calories per gram per degree Celsius (cal/g · °C), but the specific heat is the ratio of two heat capacities and is therefore dimensionless.

The heat capacity of water is 1.0 Btu/lb-°F ($= 4.2 \times 10^3$ J/kg · K), and thus the heat capacity of any material will always be numerically equal to the specific heat. Consequently, there has been the tendency to use the terms *heat capacity* and *specific heat* almost equivocally.

The specific heat of coal (Table 7.2) usually increases with its moisture content (Figure 7.1), decreases with carbon content (Figure 7.2), and increases with volatile matter content (Figure 7.3), with mineral matter content exerting somewhat less influence. The values for the specific heats of various coals fall into the general range 0.25 to 0.37, but as with other physical data, comparisons should be made only on an equal (e.g., moisture content, mineral matter content) basis.

TABLE 7.2 Specific Heat of Air-Dried Coal

	Proximate Analysis (% w/w)				Mean Specific Heat			
Source	Moisture	Volatile Matter	Fixed Carbon	Ash	28–65°C	25–130°C	25–177°C	25–227°C
West Virginia	1.8	20.4	72.4	5.4	0.261	0.288	0.301	0.314
Pennsylvania (bituminous)	1.2	34.5	58.4	5.9	0.286	0.308	0.320	0.323
Illinois	8.4	35.0	48.2	8.4	0.334			
Wyoming	11.0	38.6	40.2	10.2	0.350			
Pennsylvania (anthracite)	0.0	16.0	79.3	4.7	0.269			

Source: Baughman (1978, p. 172).

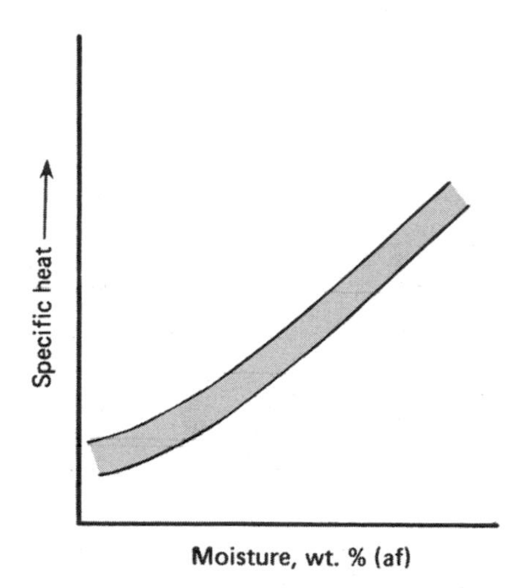

FIGURE 7.1 Variation of specific heat with moisture content. (From Baughman, 1978, p. 172.)

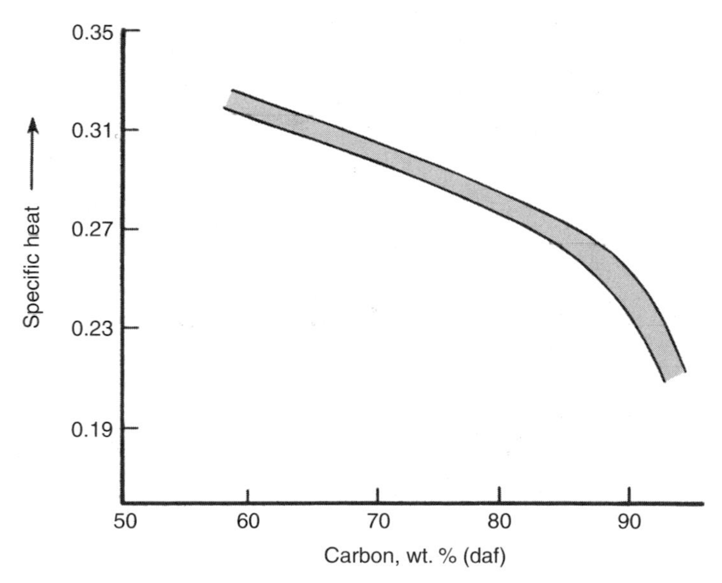

FIGURE 7.2 Variation of specific heat with carbon content. (From Baughman, 1978, p. 172.)

Estimates of the specific heat of coal have also been made on the assumption that the molecular heat of a solid material is equal to the sum of the atomic heats of the constituents (Kopp's law); the atomic heat so derived is divided by the atomic weight to give the (approximate) specific heat. Thus, from the data for various coals it has been possible to derive a formula that indicates the

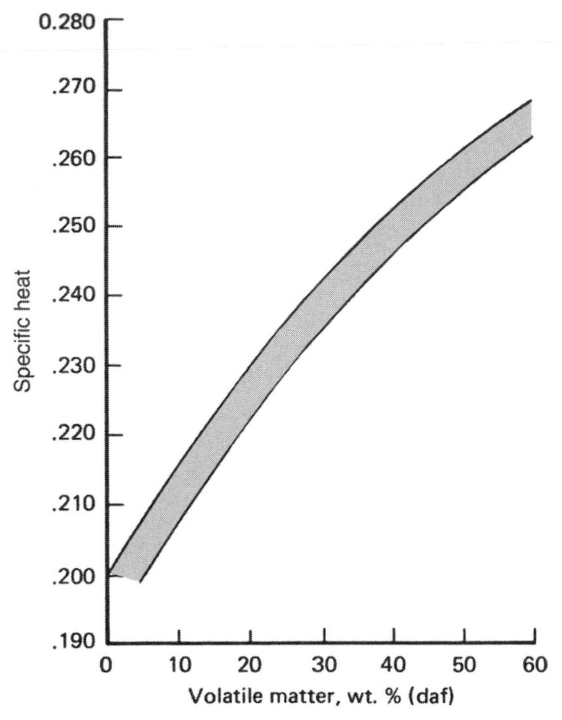

FIGURE 7.3 Variation of specific heat with volatile matter content. (From Baughman, 1978, p. 172.)

relationship between the specific heat (C_p) and the elemental analysis of coal (mmf basis):

$$C_p = 0.189\text{C} + 0.874\text{H} + 0.491\text{N} + 0.360\text{O} + 0.215\text{S} \tag{7.5}$$

where C, H, N, O, and S are the respective amounts (% w/w) of the elements in the coal.

7.3 THERMAL CONDUCTIVITY

Thermal conductivity is the rate of transfer of heat by conduction through a unit area across a unit thickness for a unit difference in temperature:

$$Q = k\frac{A(t_2 - t_1)}{d} \tag{7.6}$$

where Q is the heat, expressed as kcal/s · cm · °C or as Btu/ft-hr-°F (1 Btu/ft-hr-°F = 1.7 J/s · m · K); k the thermal conductivity; A the area; and $t_2 - t_1$ the temperature differential for the distance d (Carslaw and Jaeger, 1959).

However, the banding and bedding planes in coal (Speight, 1994, and references cited therein) can complicate the matter to such an extent that it is difficult, if not almost impossible, to determine a single value for the thermal conductivity of a particular coal. Nevertheless, it has been possible to draw certain conclusions from the data available. Thus, monolithic coal is considered to be a medium conductor of heat with the thermal conductivity of anthracite being on the order of 5 to 9×10^{-4} kcal/s · cm · °C, while the thermal conductivity of monolithic bituminous coal falls in the range 4 to 7×10^{-4} kcal/s · cm · °C. Furthermore, the thermal conductivity of pulverized coal is lower than that of the corresponding monolithic coal. For example, the thermal conductivity of pulverized bituminous coal falls into the range 2.5 to 3.5×10^{-4} kcal/s · cm · °C.

The thermal conductivity of coal generally increases with an increase in the apparent density of the coal as well as with volatile matter content, ash content, and temperature. In addition, the thermal conductivity of the coal parallel to the bedding plane appears to be higher than the thermal conductivity perpendicular to the bedding plane.

There is little information about the influence of water on the thermal conductivity of coal, but since the thermal conductivity of water is markedly higher than that of coal (about three times), the thermal conductivity of coal could be expected to increase if water is present in the coal.

7.4 PLASTIC AND AGGLUTINATING PROPERTIES

All coals undergo chemical changes when heated, but there are certain types of coal that also exhibit physical changes when subjected to the influence of heat. These particular types of coals are generally known as *caking coals*, whereas the remaining coals are referred to as *noncaking coals*.

Caking coals pass through a series of physical changes during the heating process insofar as they soften, melt, fuse, swell, and resolidify within a specific temperature range. This temperature has been called the *plastic range* of coal, and the physical changes that occur within this range have been termed the *plastic properties* (*plasticity*).

The caking tendency of coals increases with the volatile matter content of the coal and reaches a maximum in the range 25 to 35% w/w volatile matter but then tends to decrease. In addition, the caking tendency of coal is generally high in the 81 to 92% w/w carbon coals (with a maximum at 89% carbon); the caking tendency of coal also increases with hydrogen content but decreases with oxygen content and with mineral matter content.

When noncaking (nonplastic) coal is heated, the residue is pulverent and noncoherent. On the other hand, the caking coal produces a residue that is coherent and has varying degrees of friability and swelling. In the plastic range, caking coal particles have a tendency to form agglomerates (cakes) and may even adhere to surfaces of process equipment, thereby giving rise to reactor plugging problems. Thus, the plastic properties of coal are an important means of projecting and predicting how coal will behave under various process conditions as well as assisting in the selection of process equipment. For example, the plasticity of coal is beneficial in terms of the production of metallurgical coke but may have an adverse effect on the suitability of coal for conversion to liquids and gases. As an example, the Lurgi gasifier is unable to gasify caking coals adequately unless the caking properties are first nullified by prior oxidative treatment.

The plastic behavior of coal is of practical importance for semiquantitative evaluation of metallurgical coal and coal blends used in the production of coke for the steel industry. When bituminous coals are heated in the absence of air over the range 300 to 550°C (570 to 1020°F), volatile materials are released and the solid coal particles soften, to become a plasticlike mass that swells and eventually resolidifies.

7.4.1 Determination of Plastic Properties

Testing with a Gieseler plastometer (ASTM D-2639) gives a semiquantitative measurement of the plastic property, or apparent melting of coal when heated under prescribed conditions in the absence of air. The chemical nature of the constituents that account for a coal's plastic properties is not known. The material thought to be responsible for the plastic properties of coal has been removed from coal successfully by solvent extraction, leaving a nonplastic residue. Such residue has been rendered plastic by returning to it the extracts obtained by the solvent extraction. No definite relationship has been established between the amount of extract and the plastic properties of the coal.

The most common plastometer is the Gieseler plastometer (ASTM D-2639), a vertical instrument consisting of a sample holder, a stirrer with four small rabble arms attached at its lower end with the means of (1) applying a torque to the stirrer, (2) heating the sample that includes provision for controlling temperature and rate of temperature rise, and (3) measuring the rate of turning of the stirrer. In the test method, a relative measure of the plastic behavior of coal when heated under prescribed conditions is presented. This test method may be used to obtain semiquantitative values of the plastic properties of coals and blends used in carbonization and in other situations where determination of plastic behavior of coals is of practical importance, such as in the coking industry. The test for plasticity is therefore useful in studying coals and blends used in carbonization.

In the procedure (ASTM D-2639), the sample is air dried prior to preparation and the temperature should not exceed 15°C (59°F) above room temperature, and drying should not be continued to the extent that oxidation of the coal occurs and the plastic properties of the coal are not altered by oxidation. The apparatus is then immersed in the heating bath and a known torque applied to the stirrer. During the initial heating no movement of the stirrer occurs, but as the temperature is raised, the stirrer begins to rotate. With increasing temperature, the stirrer speed increases until at some point the coal resolidifies and the stirrer is halted (Figure 7.4). The plastic properties of the sample are then measured by the resistance to motion of the fluid mass in the plastometer.

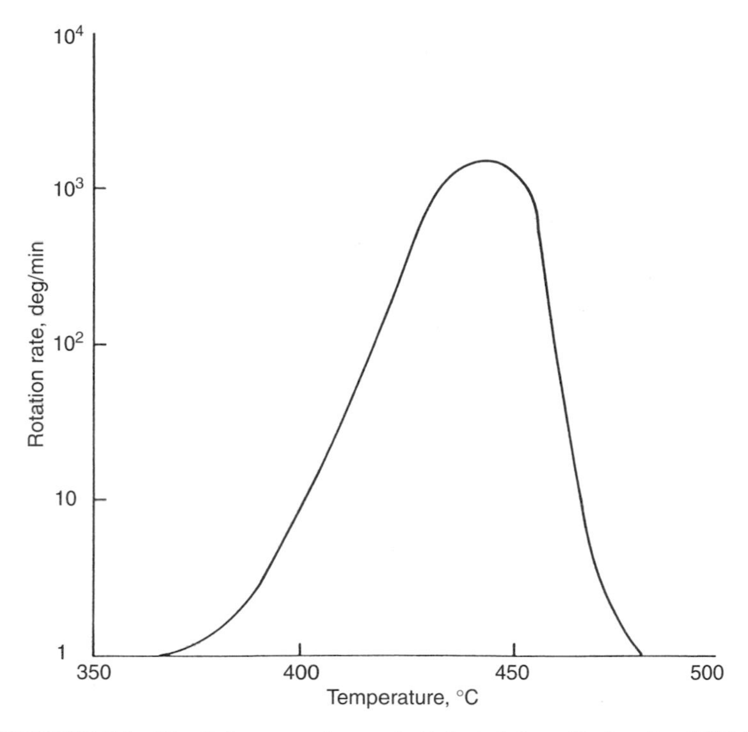

FIGURE 7.4 Plasticity curve for coal. (Adapted from Berkowitz, 1979.)

Discrepancies have been noted in results obtained with the Gieseler coal plastometer. For example, Gieseler plasticity tests performed on coking coals showed that increased maximum fluidities were associated with greater retort immersion depths (distance from center of solder bath to center of coal charge), which were reportedly caused by increased melting and volatilization rates. To achieve satisfactory interlaboratory agreement for Gieseler plasticity data, tests must be conducted using the same standard and considering oxidation effects such as those from coal storage and transport, which rapidly lower coal plasticity.

The values normally determined with the Gieseler plastometer are:

1. *Initial softening temperature*: temperature at which the dial movement reaches 1.0 dial division per minute (100 dial divisions = one complete revolution of the stirrer); may be characterized by other rates, but if so, the rate must be reported.
2. *Maximum fluid temperature*: temperature at which the dial pointer movement (stirrer revolutions) reaches the maximum rate.
3. *Solidification temperature*: temperature at which the dial pointer movement stops.
4. *Maximum fluidity*: maximum rate of dial pointer movement in dial divisions per minute.

Plastic properties are sensitive to the oxidation or weathering of coals. Maximum fluidity is lessened, and extensive oxidation may destroy the fluidity of coal completely. Samples should be tested as soon as possible after they are collected and should be stored under water or in a nonoxidizing atmosphere if there is to be a delay before they are tested. Proper packing around the stirrer in the plastometer is an important step in the measurement of plastic properties. Some coals may not pack easily due to their weathered condition or to the size consist of the sample. An excess of very fine coal makes the test sample hard to pack.

Some problems associated with the equipment used are the rate of heating, surface area of the rabble arms on the stirrer, and the manner in which torque is applied. The standard rate of heating influences values obtained in the test, with maximum fluidity being influenced the most. Heating rates higher than the standard lead to higher fluidity values, and lower rates of heating produce lower fluidity values. The plastometer must be cleaned thoroughly after each test. Frequent use and cleaning wear away the stirrer and the rabble arms, gradually decreasing their surface area. As a result, high maximum fluidity values will be obtained. When new, the rabble arms have a total surface area of 136 mm^2. When the surface area decreases to 116 mm^2 (usually after 30 to 40 tests), the rabble arms should be discarded.

Although it is possible to formulate the general stages that occur up to and during the plastic stage of coal, the exact mechanism of coal fusion is not understood completely. There seems to be little doubt that the process is concerned with the production and/or liberation of liquid tars within the coal. In terms of the elemental composition of coal, there is a relative hydrogen deficiency, but there are theories that admit to the presence of hydrogen-rich liquid (and mobile) hydrocarbons that are enclosed within the coal matrix, which are often (erroneously) called bitumen and which should not be confused with the bitumens that occur in various deposits throughout the world (Speight, 1990, 1999). The application of heat results in the liberation of these hydrocarbon liquids and forms other hydrocarbons (*thermobitumen*) by scission of hydrocarbon fragments from the coal structure, and the overall effect is the formation of a high-carbon coke and a hydrocarbon tar, the latter being responsible for the fluidity of the mass. With increased heating, the tar partly volatilizes and partly reacts to form nonfluid material ultimately leading to the coke residue.

When coal is heated in a vacuum, the plastic range is generally reduced substantially, perhaps because of the rapid evaporation of the bituminous hydrocarbons that are reputedly responsible for the fluidity of the plastic coal. Heating coal to the plastic range followed by rapid cooling yields coal with a lower softening point (if plasticized a second time), and this has been ascribed to the presence of liquid in the coal that arose from the first heating.

An additional property of coal that is worthy of mention at this time is the softening point, which is generally defined as the temperature at which the particles of coal begin to melt and become rounded. The softening point indicates the onset of the plasticity stage and is (as should be anticipated) a function of the volatile matter content of coal. For example, coal that produces 15% w/w volatile matter will have a softening point on the order of 440°C (825°F) that will decrease to a limiting value of about 340°C (645°F) for coal that produces 30% w/w volatile matter.

The material thought to be responsible for conferring the plastic properties on coal can be removed by solvent extraction to leave a nonplastic residue (Pierron and Rees, 1960). Plastic properties can be restored to the coal be recombining the solvents extracts with the insoluble residue. The fluid behavior or plasticity of coal in the temperature range 300 to 550°C (572 to 1022°F) is widely used to estimate coking characteristics and may be important for predicting coal properties with respect to coal conversion processes.

7.4.2 Data Handling and Interpretation

The primary use of plastic property data is for assessing the coking properties of coals. Maximum fluidity values are most often used in this respect, but the plastic range of coals has also been used as a guide for blending coals for carbonization. The plastic range is the temperature between the softening and the solidification temperature. Plastic property data should not be interpreted too closely. These data are probably more useful when applied to low fluid, less strongly coking coals than in assessing differences in the coking characteristics of high fluid, more strongly coking coals.

7.5 AGGLOMERATING INDEX

The agglomerating index is a grading index based on the nature of the residue from a 1-g sample of coal when heated at 950 ± 20°C (1740 ± 35°F) in volatile matter determination (ASTM D-3175).

The agglomerating index has been adopted as a requisite physical property to differentiate semianthracite from low-volatile bituminous coal and also high-volatile C bituminous coal from subbituminous A coal (Table 7.3). From the standpoint of the caking action of coal in coal-burning equipment the agglomerating index has some interest. For example, coals having indexes NAa or NAb, such as anthracite or semianthracite, certainly do not give any problems from caking, whereas those having a Cg index are, in fact, high-caking coals.

The agglomerating for agglutinating) tendency of coal may also be determined by the Roga test (ISO 335), and the Roga index (calculated from the abrasion properties when a mixture of a specific coal and anthracite is heated) is used as an indicator of the agglomerating tendencies of coals (Table 7.4).

7.6 FREE-SWELLING INDEX

The free-swelling index (FSI) is a measure of the increase in volume of coal when heated under specified conditions (ASTM D-720; ISO 335).

TABLE 7.3 Agglomerating and Caking Properties of Coal

Designation		
Class[a]	Group	Appearance of Residue
Nonagglomerating (button shows no swelling or cell structure and will not support a 500-g weight without pulverizing)	NA, nonagglomerate	NAa, noncoherent residue Nab, coke cutton shows no swelling or cell structure and after careful removal from the crucible will pulverize under a 500-g weight carefully lowered onto the button
	A, agglomerate (button dull black, sintered, shows no swelling, or cell structure will support a 500-g weight without pulverizing)	Aw, weak agglomerate (buttons come out of crucible in more than one piece) Af, firm agglomerate (buttons come out of crucible in one piece)
Agglomerating (button shows swelling or cell structure or will support a 500-g weight without pulverizing)	C, caking (button shows swelling or cell structure)	Cp, poor caking (button shows slight swelling with small cells, has slight gray luster) Cf, fair caking (button shows medium swelling and good cell structure; has characteristic metallic luster) Cg, good caking (button shows strong swelling and pronounced cell structure with numerous large cells and cavities; has characteristic metallic luster)

[a] Agglomerating index: Coals that in the volatile matter determination produce either an agglomerate button that will support a 500-g weight without pulverizing or a button showing swelling or cell structure shall be considered agglomerating from the standpoint of classification.

TABLE 7.4 Plastic Properties of Coal

Coal Type	Swelling Index	Dilation (%)	Roga Index
Noncaking	0	0	0–5
Weakly caking	1–2	0	5–20
Medium caking	2–4	0–40	20–50
Strongly caking	>4	>50	>50

TABLE 7.5 Free-Swelling Index of Coal

Rank	Coal	Free-Swelling Index
High-volatile C	Illinois No. 6	3.5
High-volatile B	Illinois No. 6	4.5
High-volatile B	Illinois No. 5	3.0
High-volatile A	Illinois No. 5	5.5
High-volatile A	Eastern	6.0–7.5
Medium-volatile	Eastern	8.5
Low-volatile	Eastern	8.5–9.0

Source: Baughman (1978, p. 176).

The results from a test may also be used as an indication of the caking characteristics of the coal when it is burned as a fuel. The volume increase can be associated with the plastic properties of coal; coals that do not exhibit plastic properties when heated do not show free swelling. It is believed that gas formed by thermal decomposition while the coal is in a plastic or semifluid condition is responsible for the swelling. The amount of swelling depends on the fluidity of the plastic coal, the thickness of bubble walls formed by the gas, and interfacial tension between the fluid and solid particles in the coal. When these factors cause more gas to be trapped, greater swelling of the coal occurs.

The free-swelling index of bituminous coals generally increases with an increase in rank (Table 7.5). Values for individual coals within a rank may vary considerably. The values for the lower-rank coals are normally less than those for bituminous coals, whereas anthracite does not fuse and shows no swelling value.

7.6.1 Determination of the Free-Swelling Index

The test method (ASTM D-720) is a small-scale test for obtaining information regarding the free-swelling properties of coal. The results may be used as an indication of the caking characteristic of the coal when burned as a fuel. This test is not recommended as a method for the determination of expansion of coals in coke ovens.

In the test method (ASTM D-720), a weight amount (approximately 1 g) of the sample is placed in a translucent silica crucible with a prescribed size and shape, and the sample is leveled in the crucible by light tapping on a hard surface. The cold crucible is then lowered into a special furnace and heated to 800 ± 10°C (1472 ± 18°F) in 1 minute and 820 ± 5°C (1508 ± 9°F) in 1 minute. The test can be made with either gas or electric heating. The button formed in the crucible is then compared to a chart of standard profiles and corresponding swelling index numbers. Three to five buttons are made for each sample, and the average of the profile numbers is taken as the free-swelling index. The shape, or profile, of the buttons (Table 7.5) determines the free-swelling index of the coal. Anthracite does not usually fuse or exhibit a free-swelling index, whereas the free-swelling index of bituminous coal will increase as the rank increases from high-volatile C bituminous coal to the low-volatile bituminous coal.

Some problems associated with the method are the proper heating rate, oxidation or weathering of the coal sample, and an excess of fine coal in the analysis sample. Failure to achieve the proper temperature in the furnace or, more important, the proper heating rate for the sample in the crucible leads to unreliable results. Uneven heat distribution along the walls of the crucible may also cause erratic results. Oxidation or weathering of the coal sample leads to a low free-swelling index. To minimize oxidation and the effects on the free-swelling index, samples should be tested as soon as possible after they are collected and prepared. If oxidation of the coal is suspected, the test should be repeated on a known fresh sample of the same coal.

The size consist of the analysis sample may influence the free-swelling index values of some coals. There is evidence that for many coals, an excess of fine coal (100 to 200 mesh) may cause FSI values to be as much as two index numbers high. The amount of fine coal in the analysis sample should be kept at a minimum for this test (and others). Reducing the coal from a large particle size to a small particle size in one step tends to produce a high concentration of fine coal. The reduction of coal samples should be done in an appropriate number of steps to avoid this.

In another test method (ASTM D-5515), a dilatometer is used to measure the swelling of bituminous coal. The test method is limited in applicability to coal that has a free swelling index ≤ 1 (ASTM D-720). The principle of this test method is that the final volume of char obtained at the conclusion of a standard dilatation test is dependent on the mass of coal in the coal pencil and on the radius of the retort tube. This test method incorporates a procedure that determines the mass of air-dried coal in the coal pencil, provides a means to measure the average retort tube radii; and employs a means to report coal expansion on an air-dried coal weight basis. The value of the dilatation properties of coals may be used to predict or explain the behavior of a coal or blends during carbonization or in other processes, such as gasification and combustion.

Other test methods used to determine the swelling properties of bituminous coals include the Ruhr test method (ISO 8264) and Audibert–Arnu (ISO 349) test method. However, these two test methods provide consistently different values for percent dilatation and percent contraction. Percent contraction and dilatation values obtained using the Audibert–Arnu test method are higher and lower, respectively, than those obtained using the Ruhr test method. These differences have been attributed to trimming the length of the coal pencil from different ends. The Audibert–Arnu test method specifies that the wider end of the coal pencil be trimmed, whereas the Ruhr test method specifies that the narrower end of the coal pencil be trimmed.

The Roga test (ISO 335) measures mechanical strength rather than size profiles of coke buttons; another test (ISO 501) gives a crucible swelling number of coal. The nature of the volume increase is associated with the plastic properties of coal (Loison et al., 1963), and as might be anticipated, coals that do not exhibit plastic properties when heated do not, therefore, exhibit free swelling. Although this relationship between free-swelling and plastic properties may be quite complex, it is presumed that when the coal is in a plastic (or semifluid) condition, the gas bubbles formed as a part of the thermal decomposition process within the fluid material cause the swelling phenomenon, which, in turn, is influenced by the thickness of the bubble walls, the fluidity of the coal, and the interfacial tension between the fluid material and the solid particles that are presumed to be present under the test conditions.

7.6.2 Data Handling and Interpretation

The test for the free-swelling index (FSI) is an empirical one, and FSI values can be used to indicate the coking characteristics of coal when burned as a fuel. However, these values are not reliable enough for use as parameters in a classification system. Free-swelling index values have been considered useful as an indication of the tendency of coals to form objectionable "coke trees" when burned in certain types of equipment, particularly equipment with underfeed stokers. The decline in the use of underfeed stokers in coal-burning equipment along with adjustments of combustion conditions have minimized the problems due to coke tree formation. The use of free-swelling index test data for help in solving this problem has also declined.

In general terms, the free-swelling index of coal generally increases with an increase in rank (Rees, 1966), but the values for individual coals within a rank may vary considerably. The values for the lower-rank coals are normally less than those for bituminous coals; anthracite does not fuse and shows no swelling value. Furthermore, a coal exhibiting a free-swelling index of 2, or less, will probably not be a good coking coal, whereas a coal having a free-swelling index of 4 or more may have good coking properties. The free-swelling index can also be used as an indication of the extent of oxidation or weathering of coals. However, these are not as sensitive to weathering as calorific values.

7.7 ASH FUSIBILITY

Coal ash (Chapter 3) is the noncombustible residue that remains after all the combustible material has been burned. It is a complex mixture that results from chemical changes that take place in the components of the coal mineral matter. The composition of coal ash varies extensively just as the composition of coal mineral matter varies.

The ash fusibility test method (ASTM D-1857) is designed to simulate as closely as possible the behavior of coal ash when it is heated in contact with either a reducing or an oxidizing atmosphere. The test is intended to provide information on the fusion characteristics of the ash. It gives an approximation of the temperatures at which the ash remaining after the combustion of coal will sinter, melt, and flow. *Sintering* is the process by which the solid ash particles weld together without melting. The temperature points are measured by observation of the behavior of triangular pyramids (cones) prepared from coal ash when heated at a specified rate in a controlled atmosphere. The critical temperature points are as follows:

1. *Initial deformation temperature* (IT): temperature at which the first rounding of the apex of the cone occurs
2. *Softening temperature* (ST): temperature at which the cone has fused down to a spherical lump in which the height is equal to the width of the base
3. *Hemispherical temperature* (HT): temperature at which the cone has fused down to a hemispherical lump, at which point the height is one-half the width of the base
4. *Fluid temperature* (FT): temperature at which the fused mass has spread out in a nearly flat layer with a maximum height of J_6 inches.

When determining the initial deformation temperature, shrinkage or warping of a cone is ignored if the tip remains sharp (Figure 7.5).

7.7.1 Determination of Ash Fusibility

The test method for determining the fusibility of coal ash (ASTM D-1857) covers the observation of the temperatures at which triangular pyramids (cones) prepared from coal and coke ash attain and pass through certain defined stages of fusing and flow when heated at a specified rate in controlled, mildly reducing, and where desired, oxidizing atmospheres.

In the test method (ASTM D-1857; ISO 540), coal passing a number 60 (250 μm) sieve (analysis sample prepared in accordance with method ASTM D-2013) is heated gradually to a temperature of 800 to 900°C (1472 to 1652°F)

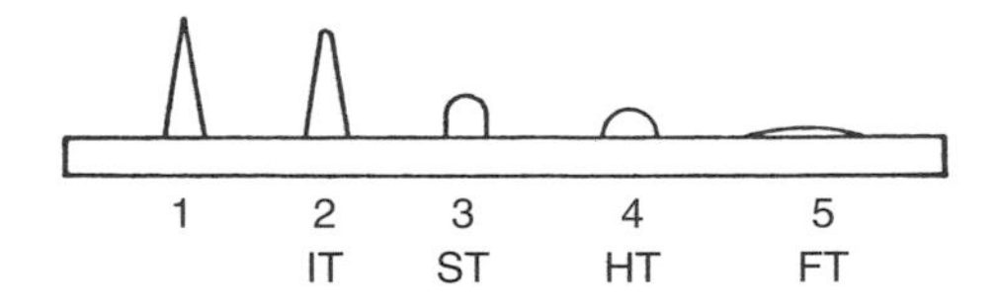

FIGURE 7.5 Critical temperature points. (From ASTM, 2004.)

to remove most of the combustible material. The ash is ground in an agate mortar to pass a number 200 (75 μm) sieve, spread on a suitable dish, and ignited in a stream of oxygen for approximately 1 hour at 800 to 850°C (1472 to 1562°F). Enough coal is used to produce 3 to 5 g of ash. The ash is mixed thoroughly and moistened with a few drops of dextrin binder and worked into a stiff plastic mass. The mass is then formed into a cone using a cone mold. The cones are dried, mounted on a refractory base, and heated at a specified rate in a gas-fired or electrically heated furnace under either oxidizing or reducing conditions.

In gas-fired furnaces, regulating the ratio of air to combustible gas controls the atmosphere. For reducing conditions an excess of gas over air is maintained, and for oxidizing conditions an excess of air over gas is maintained. Hydrogen, hydrocarbons, and carbon monoxide produce a reducing atmosphere, while oxygen, carbon dioxide, and water vapor are considered to be oxidizing gases. For a mildly reducing atmosphere, the ratio by volume of reducing gases to oxidizing gases must be maintained between the limits of 20–80 and 80–20 on a nitrogen-free basis. In a gas-fired furnace, this ratio may be difficult to achieve at high temperatures while maintaining the required temperature rise. For an oxidizing atmosphere, the volume of reducing gases present must not exceed 10%.

In electrically heated furnaces, a mixture of 60% v/v carbon monoxide and 40% v/v ± 5% v/v carbon dioxide produces a reducing atmosphere in the furnace. A regulated stream of air produces an oxidizing atmosphere.

Proper control of the atmosphere surrounding the test specimen is probably the greatest issue encountered in determining ash fusibility, particularly when a reducing atmosphere is used. A mildly reducing atmosphere is specified since it is believed that this more closely approximates conditions existing in fire beds when coal is burned in several types of combustion equipment. Lower softening temperature values are obtained with a mildly reducing atmosphere than in either strongly reducing or oxidizing atmospheres. With a mildly reducing atmosphere, the iron in the ash is present predominantly in the ferrous state, whereas in a strong reducing atmosphere, some of the iron may be in the metallic state. In an oxidizing atmosphere the iron is in the ferric state. Both ferric iron and metallic iron increase the refractory quality of the ash, resulting in higher fusion temperatures. Softening temperature values may vary by as much as 150 to 200°C (302 to 392°F), depending on the atmosphere in which the test is made.

In preparing ash for the fusibility test, it is important that the coal be spread out in a thin layer and that adequate circulation of air be maintained during burning. All iron must be converted to the ferric state, and all combustible matter must be removed. A low initial heating temperature and a slow heating rate tend to minimize the retention of sulfur as sulfates in the ash. Following the burning in air, pulverizing the ash and burning it in oxygen will ensure complete conversion of iron to the ferric state and that all combustible material is burned.

7.7.2 Data Handling and Interpretation

Ash fusibility values are often specified in coal contracts because they are believed to be a measure of the tendency of coal ash to form clinkers. Softening temperatures probably are used most often for this purpose. For instance, if it is desirable to have the ash fuse into a large clinker that could easily be removed, coal with a softening temperature low enough to allow the ash to fuse would be chosen. However, the ash should not soften at too low a temperature, as it may become fluid enough to run through the fire bed and solidify below it, making the ash harder to remove. Coal with high softening temperatures produces ash with relatively small particle size rather than fused masses. Initial deformation and fluid temperatures may also be useful, depending on the type of combustion equipment to be used for burning coal and the manner in which the ash is to be removed.

In practice, types of burning equipment, rate of burning, temperature and thickness of the fire bed, distribution of ash-forming minerals in the coal, and viscosity of the molten ash may influence ash behavior more than do the laboratory-determined ash fusibility characteristics. The correlation of the laboratory test with the actual utilization of coal is only approximate, due to the relative homogeneity of the laboratory test sample compared to the heterogeneous mixture of ash that occurs when coal is burned. Conditions that exist during the combustion of coal are so complex that they are impossible to duplicate completely in a small-scale laboratory test. Therefore, the test should be considered only as an empirical one, and the data should be considered qualitative and should not be overinterpreted.

7.8 THERMAL CONDUCTIVITY (DIFFUSIVITY AND EXPANSION)

Coal is a low-to-medium conductor of heat, with thermal conductivity values ranging from about 3×10^{-4} to 9×10^{-4} cal/s · cm · °C (Speight, 1994, and references cited therein). The thermal conductivity λ of coal is related to thermal diffusivity (α) and the heat capacity (C_p) by

$$\alpha = \frac{\lambda}{\rho C_p} \tag{7.7}$$

where ρ is the density. Thermal conductivity is the time rate of heat transfer by conduction across a unit area of unit thickness for a unit temperature gradient. Thermal diffusivity is a measure of the rate at which a temperature wave travels.

Steady-state periodic heating and unsteady-state methods can be applied to measure the thermal conductivity and diffusivity of coal. Methods such as the compound bar method and calorimetry have been replaced by transient hot-wire/line heat source, and transient hot plate methods that allow very rapid and independent measurements of α and λ. In fact, such methods offer the additional advantage of measuring these properties not only for monolithic samples but also for coal aggregates and powders under conditions similar to those encountered in coal utilization systems.

Due to contributions from mineral matter (pyrite being of great importance), no systematic correlation exists between coal rank and thermal conductivity/diffusivity values. On the other hand, a reasonably good correlation has been observed with uncorrected coal densities (Speight, 1994, and references cited therein). These properties also exhibit anisotropy; thermal conductivity and diffusivity are highest for samples parallel to the bedding plane.

Thermal conductivity increases with increasing apparent density, volatile matter, ash, and mineral matter content. Due to the high porosity of coal, thermal conductivity is also strongly dependent on the nature of gas, vapor, or fluid in the pores, even for monolithic samples (van Krevelen, 1961). Moisture has a similar effect and increases the thermal conductivity of coal since its thermal conductivity value is approximately three times higher than that of dry coal (Speight, 1994, and references cited therein). However, the thermal diffusivity of coal is practically unaffected by moisture since the λ/C_p value is not essentially changed by moisture.

No overall model applicable to the prediction of thermal conductivity/diffusivity values is available, but assuming the presence of additive contributions from the elements in coal, the following correlation has been proposed:

$$\lambda^{-1}(\mathrm{mK/W}) = \frac{\mathrm{C}}{1.47} + \frac{\mathrm{H}}{0.0118}\left(\frac{273}{T}\right)^{0.5} \tag{7.8}$$

where C and H are the mass fractions of carbon and hydrogen, and T is the absolute temperature in kelvin; this equation yields acceptable predictions for monolithic bituminous coal.

The thermal conductivity of crushed/pulverized coal is lower than that of monolithic coal. The model (above) for such two-phase systems consisting of a discontinuous phase of coal particles, and a continuous gas phase is in broad agreement with literature data and predicts that the thermal conductivity of monolithic bituminous coals is more than twice the value of their aggregates.

The linear thermal expansion coefficient shows the greatest increase in temperature for bituminous coals. The values for the linear thermal expansion coefficient are less than $33 \times 10^{-6}\ ^{\circ}\mathrm{C}^{-1}$ in the 30 to 330°C (86 to 626°F) range (van Krevelen, 1961). For anthracite, the linear thermal expansion coefficient changes very little with temperature and is accompanied by a pronounced anisotropy effect. The values for the linear thermal expansion coefficient are about twice as high for coal perpendicular to the bedding plane than for coal parallel to the bedding plane (van Krevelen, 1961).

REFERENCES

ASTM. 2004. *Annual Book of ASTM Standards*, Vol. 05.06. American Society for Testing and Materials, West Conshohocken, PA. Specifically:

ASTM C-351. Standard Test Method for Mean Specific Heat of Thermal Insulation.

ASTM D-121. Terminology of Coal and Coke.

ASTM D-388. Standard Classification of Coals by Rank.

ASTM D-720. Test Method for Free-Swelling Index of Coal.

ASTM D-1857. Standard Test Method for Fusibility of Coal and Coke Ash.

ASTM D-2013. Standard Practice of Preparing Coal Samples for Analysis.

ASTM D-2015. Test Method for Gross Calorific Value of Coal and Coke by the Adiabatic Bomb Calorimeter.

ASTM D-2639. Test Method for Plastic Properties of Coal by the Constant-Torque Gieseler Plastometer.

ASTM D-3175. Standard Test Method for Volatile Matter in the Analysis Sample of Coal and Coke.

ASTM D-3286. Test Method for Gross Calorific Value of Coal and Coke by the Isoperibol Bomb Calorimeter.

ASTM D-5515. Standard Test Method for Determination of the Swelling Properties of Bituminous Coal Using a Dilatometer.

ASTM D-5865. Standard Test Method for Gross Calorific Value of Coal and Coke.

Baughman, G. L. 1978. *Synthetic Fuels Data Handbook*. Cameron Engineers, Denver, CO.

Berkowitz, N. 1979. *An Introduction to Coal Technology*, Academic Press, San Diego, CA.

Carslaw, H. S., and Jaeger, J. C. 1959. *Conduction of Heat in Solids*, 2nd ed. Oxford University Press, Oxford, p. 189.

Chan, M. -L., Parkyns, N. D., and Thomas, K. M. 1991. *Fuel*, 70:447.

ISO. 2004. *Standard Test Methods for Coal Analysis*. International Organization for Standardization, Geneva, Switzerland. Specifically:

ISO 335. Determination of Coking Power of Hard Coal: Roga Test.

ISO 349. Hard Coal: Audibert–Arnu Dilatometer Test.

ISO 501. Determination of Crucible Swelling Number.

ISO 540. Determination of the Fusibility of Ash: High-Temperature Tube Method.

ISO 1928. Determination of Gross Calorific Value and Calculation of Net Calorific Value.

ISO 8264. Determination of the Swelling Properties of Hard Coal Using a Dilatometer.

Loison, R., Peytavy, A., Boyer, A. F., and Grillot, R. 1963. In *Chemistry of Coal Utilization*, Suppl. Vol., H. H. Lowry (Editor). Wiley, Hoboken, NJ, Chap. 4.

Munoz-Guillena, M. J., Linares-Solano, A., and Salinas-Martinez de Lecea, C. 1992. *Fuel*, 71:579.

Pierron, E. D., and Rees, O. W. 1960. *Solvent Extract and the Plastic Properties of Coal*. Circular 288. Illinois State Geological Survey, Urbana, IL.

Rees, O. W. 1966. *Chemistry, Uses, and Limitations of Coal Analysis*. Report of Investigations 220. Illinois State Geological Survey, Urbana, IL.

Selvig, W. A. 1945. In *Chemistry of Coal Utilization*. H.H. Lowry (Editor). Wiley, Hoboken, NJ, Chap. 4.

Speight, J. G. 1990. In *Fuel Science and Technology Handbook*, J. G. Speight (Editor). Marcel Dekker, New York, Chaps. 2 and 12.

Speight, J. G. 1994. *The Chemistry and Technology of Coal*, 2nd ed. Marcel Dekker, New York.

Speight, J. G. 1999. *The Chemistry and Technology of Petroleum*, 3rd ed. Marcel Dekker, New York.

van Krevelin, D. W. 1961, *Coal*. Elsevier, Amsterdam.

8 Mechanical Properties

In contrast to proximate analysis, ultimate (elemental) analysis, and the physical properties, the mechanical properties of coal have been little used and are not often reported in the scientific literature. This is a serious omission because these properties are of importance and should be of consideration in predicting coal behavior during mining, handling, and preparation.

For example, the mechanical properties of coal are of value as a means of predicting the strength of coal and its behavior in mines when the strength of coal pillars and stability of coal faces are extremely important factors. The mechanical properties of coal are also of value in areas such as coal winning (for the design and operation of cutting machinery), comminuting (design and/or selection of mills), storage (flow properties, failure under shear), and handling (shatter and abrasion during transport), as well as in many other facets of coal technology (Yancey and Geer, 1945; van Krevelen, 1957; Brown and Hiorns, 1963; Trollope et al., 1965; Evans and Allardice, 1978).

8.1 STRENGTH

Several methods of estimating relative strength or friability or grindability utilize a porcelain jar mill in which each coal may be ground for, say, revolutions, and the amount of new surface is estimated from screen analyses of the feed and of the ground product. Coals are then rated in grindability by comparing the amount of new surface found in the test with that obtained for a standard coal.

The only standard test method that is available is actually a test method for determining coke reactivity and coke strength after reaction (ASTM D-5341). This test method describes the equipment and techniques used for determining lump coke reactivity in carbon dioxide (CO_2) gas at elevated temperatures and its strength after reaction in carbon dioxide gas by tumbling in a cylindrical chamber.

In the test method as written, a sample of dried coke of designated origin and size is reacted [at $1100 \pm 5°C$ ($2012 \pm 9°F$)] with carbon dioxide gas in a retort at a specified elevated temperature for a specified length of time. Two indexes, the coke reactivity index (CRI) and coke strength after reaction (CSR), are determined using the reacted coke residue. The weight loss after reaction determines the coke reactivity index. The weight retained after sieving the tumbled reacted coke in a designated number of revolutions over a designated turning

Handbook of Coal Analysis, by James G. Speight
ISBN 0-471-52273-2

rate determines the coke strength after reaction. The test method is significant for coke because when coke lumps descend in the blast furnace, they are subjected to reaction with countercurrent carbon dioxide and to abrasion as they rub together and against the walls of the furnace. These concurrent processes weaken physically and react chemically with the coke lumps, producing an excess of fines that can decrease burden permeability and result in increased coke rates and lost hot metal production. This test method is designed to indirectly measure this behavior of coke in the blast furnace. With some modification, depending on the coal and the data required, the test method can be applied to determining the strength of coal.

The most common strength characteristics of coal involve investigations of the compressive strength, and this interest arises because of the relationship to the strength of coal pillars used for roof support in mining operations. Thus, it is not surprising that many tests have arisen as a means of measuring the compressive strength of coal. The tests vary with the nature of the investigations and the coal under test to such an extent that the methodology can be questioned if it is not applied in a precise and logical manner adequate to the coal to be tested and to the task.

For example, anthracite has been tested in the form of prisms (having 2 × 2 in., 5 × 5 cm bases that were cut parallel to the bedding planes of the coal with heights of 1, 2, and 4 in.) by the application of compression perpendicular to the bedding planes. The crushing strengths of the 1-in.-high prisms ranged from 3200 to 10,900 psi and averaged approximately 6000 psi. Furthermore, it has also been shown that the compressive strength of coal is inversely proportional to the square root of the height of the specimen under test. However, the lateral dimension also influences the strength of bituminous coal specimen, the smaller specimens showing greater strength than the larger (Table 8.1), which can be attributed to the presence in the larger specimen of fracture planes or cleats. In fact, it is the smaller samples that present a more accurate indication of the strength of the coal. A factor that probably contributes to the lower strength of large blocks of coal tested in the laboratory is the difficulty of mining large blocks and transporting them from the mine to the laboratory without imposing strains that start disintegration along the cleats.

TABLE 8.1 Variation of Compressive Strength with Size

Size	Average Maximum Load
2.5–4-in. cubes	2486
7–8-in. cubes	2170
10–12-in. cubes	2008
12 × 12 × 18 in. high	1152
Approx. 30-in. cube	817
Approx. 54-in. cube	306

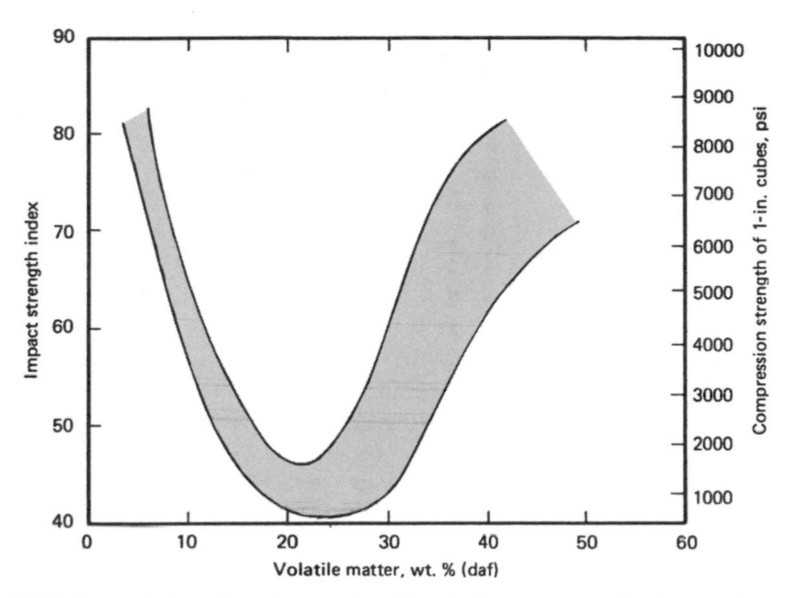

FIGURE 8.1 Variation of coal strength with volatile matter production. (Adapted from Brown and Hiorns, 1963, p. 131.)

TABLE 8.2 Variation of Compressive Strength with Rank

Coal Rank	Compressive Strength (psi)
Anthracite	2370
	3390
	2000
	1740
Bituminous	310–2490
	500
Lignite	2190–6560

The variation of compressive strength with rank of coals has been noted and a plot of strength against volatile matter shows the customary minimum to be 20 to 25% dry, ash-free volatile matter (Table 8.2 and Figure 8.1) for compression both perpendicular and parallel to the bedding plane.

8.2 HARDNESS

Considerable information is available on the properties of coal that relate to its hardness, such as friability and grindability, but little is known of the hardness of coal as an intrinsic property. However, hardness can also be equated with friability, and a test method is available to determine the friability of coal (ASTM D-441; see below) that all but mirrors the test method for determining the relative measure of the resistance of coke to degradation. In this test method (ASTM D-3402), a procedure for obtaining a relative measure of the resistance to degradation of coke by impact and abrasion is described. In the method as written, a sample of dry coke of designated size is tumbled in a rotating drum at a specified turning rate for a specified number of revolutions. Two indexes of its strength, the stability factor and the hardness factor, are determined by sieve analysis of the coke after treatment.

This test method can be applied to coal, remembering the reason for which the method was originally designed, and the data are applicable to coal hardness. Application of the test to coal, usually being a softer material than coke, will require some adjustment of the experimental conditions, but the data can be useful in determining the hardness of coal and its possible degree of degradation during shipping and handling. These processes subject the coke to impact and abrasion and the test method is a relative measure of the resistance of coal to breakage when subjected to these degradation processes.

The scratch hardness of coal can be determined by measuring the load on a pyramidal steel point required to make a scratch 100 μm in width on the polished surface of a specimen. The scratch hardness of anthracite is approximately six times that of soft coal, whereas pyrite is almost 20 times as hard as soft coal (Table 8.3). Similar data were noted for anthracite and cannel coal, but durain, the reputedly hard component of coal, was found to be only slightly harder than vitrain and cannel coal.

Although the resistance of coals to abrasion may have little apparent commercial significance, the abrasiveness of coal is, on the other hand, a factor of considerable importance. Thus, the wear of grinding elements due to the abrasive action of coal results in maintenance charges that constitute one of the major items

TABLE 8.3 Scratch Hardness of Coal

Material	Scratch Hardness Relative to Barnsley Soft Coal
Anthracite, Great Mountain	1.70
Anthracite, Red Vein	1.75
Welsh steam	0.29
Barnsley hards	0.85
Barnsley softs	1.00
Illinois coal	1.10
Cannel	0.92
Carbonaceous shale	0.69
Shale	0.32
Pyrite	5.71
Calcite	1.92

in the cost of grinding coal for use as pulverized fuel. Moreover, as coals vary widely in abrasiveness, this factor must be considered when coals are selected for pulverized fuel plants. A standardized, simple laboratory method of evaluating the abrasiveness of coal would assist, like the standard grindability test now available, in the selection of coals suitable for use in pulverized form. Actually, the abrasiveness of coal may be determined more by the nature of its associated impurities than by the nature of the coal substance. For example, pyrite is 20 times harder than coal, and the individual grains of sandstone, another common impurity in coal, also are hard and abrasive.

8.3 FRIABILITY

One measure of the strength of coal is its ability to withstand degradation in size on handling. The tendency toward breakage on handling (friability) depends on toughness, elasticity, and fracture characteristics as well as on strength, but despite this fact, the friability test is the measure of coal strength used most frequently.

Friability is of interest primarily because friable coals yield smaller proportions of the coarse sizes that may (depending on use) be more desirable, and there may also be an increased amount of surface in the friable coals. This surface allows more rapid oxidation; hence, conditions are more favorable for spontaneous ignition, loss in coking quality in coking coals, and other changes that accompany oxidation. These economic aspects of the friability of coal have provided the incentive toward development of laboratory friability tests.

The tumbler test for measuring coal friability (ASTM D-441) covers determination of the relative friability of a particular size of sized coal. It affords a means of measuring the liability of coal to break into smaller pieces when subjected to repeated handling at the mine or, subsequently, by the distributor or by the consumer. This test method is useful for determining the similarity of coals in respect to friability rather than for determining values within narrow limits to emphasize their dissimilarity. This test method may also serve to indicate the relative extent to which sized coals will suffer size degradation in certain mechanical feed devices. This test method may be used for differentiating between certain ranks and grades of coal, and therefore the test method is also useful for application to coal classification.

The method employs a cylindrical porcelain jar mill fitted with three lifters that assist in tumbling the coal. A sample (usually 1000 g) of sized coal is tumbled in the mill for a specified time at a specified number of revolutions per minute. The coal is then removed and screened and the friability is reported as the percentage reduction in the average particle size during the test. For example, if the average particle size of the tumbled coal was 75% that of the original sample, the friability would be 25%. As with several other tests, the test parameters can be adjusted to suit the purpose of the investigation, but the precise parameters must be reported with the data.

Friability (or hardness) can also be determined by the use of a drop shatter test ASTM D-440) that covers the determination of the relative size stability and its complement, the friability, of sized coal. The test method affords a means of indicating the ability of coal to withstand breakage when subjected to handling at the mine and during transit. The test method is useful for determining the similarity of coals in respect to size stability and friability rather than for determining values within narrow limits in order to emphasize their dissimilarity. This test method is considered applicable for testing a selected single size of different coals, for testing different single sizes of the same coal, and for mixed sizes of the same or different coals. This test method appears best suited for measuring the relative resistance to breakage of the larger sizes of coal when handled in thin layers, such as from loader to mine car, from loading boom to railroad car, and from shovel to chute. Although the method may not be so well adapted for measuring the potential for coal breakage when handled in mass, as in unloading open-bottom cars or emptying bins, the test method will also serve to indicate the relative size stability of composite sizes of coal, where in commercial handling, the smaller pieces have a cushioning effect, which tends to lessen breakage of the larger pieces of coal. A similar drop shatter test has also been described for determining the friability of coke (ASTM D-3038).

In the drop shatter test for coal (ASTM D-440), a 50-lb sample of coal [4.5 to 7.6 cm (2 to 3 in.)] is dropped twice from a drop-bottom box onto a steel plate 6 ft below the box. The material shattered by the two drops is then screened (using specified screens) and the average particle size is determined. The average size of the material, expressed as a percentage of the size of the original sample, is termed the *sizability*, and its complement, the percentage of reduction in average particle size, is termed the *percentage friability*. Provision is made for testing sizes other than that stipulated for the standard test to permit comparison of different sizes of the same coal.

Attempts have been made to correlate the friability of coal with rank (Table 8.4). Thus, lignite saturated with moisture was found to be the least friable, and friability increased with coal rank to a maximum for coals of the low-volatile bituminous coal. The friability of anthracites is comparable with that of

TABLE 8.4 Variation of Friability with Rank

Coal Rank	Number of Tests	Friability (%)
Anthracite	36	33
Bituminous (low-volatile)	27	70
Bituminous	87	43
Subbituminous A	40	30
Subbituminous B	29	20
Lignite	16	12

Source: Yancey and Geer (1945).

subbituminous coal; both are stronger than bituminous coal and decidedly more resistant to breakage than some of the extremely friable semibituminous coals.

The relationship between the friability of coal and its rank has a bearing on its tendency to undergo spontaneous heating and ignition (Chakravorty, 1984; Chakravorty and Karr, 1986). The friable, low-volatile coals, because of their high rank, do not oxidize readily despite the excessive fines and the attendant increased surface they produce on handling. Coals of somewhat lower rank, which oxidize more readily, usually are relatively nonfriable; hence they resist degradation in size, with its accompanying increase in the amount of surface exposed to oxidation. But above all, the primary factor in coal stockpile instability is unquestionably oxidation by atmospheric oxygen, whereas the role of any secondary factors such as friability is to exacerbate the primary oxidation effect (Jones and Vais, 1991).

8.4 GRINDABILITY

Grindability is an index of the relative ease with which a coal may be pulverized in comparison with coals chosen as standards. The Hardgrove method (ASTM D-409) has been accepted as the standard method of grindability of coal by the Hardgrove machine. There is also a test method to determine the Hardgrove grindability index of petroleum coke (ASTM D-5003) that might be applicable to extremely hard coal if the usual test method (ASTM D-409) does not produce suitable reproducibility and repeatability. Suitable modifications of the test method for petroleum coke might be required.

The Hardgrove grindability test method covers the determination of the relative grindability or ease of pulverization of coal in comparison with coal chosen as the standard. A prepared and sized sample receives a definite amount of grinding energy in a miniature pulverizer, and the size consist of the pulverized product is determined by sieving. The resulting size distribution is used to produce an index relative to the ease of grinding.

High-volatile bituminous coal, subbituminous coal, and lignite can undergo physical change as the natural or seam moisture is released during handling and preparation. This change is often sufficient to alter the grindability characteristics that will be reported when tested in the laboratory and could produce different indexes, depending on the conditions of drying and the moisture level, causing inconsistencies in repeatability and reproducibility of the data for these coals.

The test for grindability (ASTM D-409; ISO 5074; Hardgrove, 1932; Edwards et al., 1980) utilizes a ball-and-ring type of mill in which a sample of closely sized coal is ground for specified number (usually, 60) revolutions, after which the ground product is sieved and the grindability index is calculated from the amount of undersize produced using a calibration chart (Table 8.5 and Figure 8.2). The results are converted into the equivalent Hardgrove grindability index. High grindability index numbers indicate easy-to-grind coals. Each Hardgrove machine should be calibrated by use of standard reference samples of coal with grindability indexes of approximately 40, 60, 80, and 110. These numbers are based on

TABLE 8.5 Hardgrove Grindability Index of Coal

State	County	Bed	Mine	Hardgrove Grindability Index
Alabama	Walker	Black Creek	Drummond	46
Colorado	Fremont		Monarch No. 4	46
	Mesa		Cameo	47
Illinois	Fulton	No. 2	Sun Spot	53
	Stark	No. 6	Allendale	61
	Williamson	No. 6	Utility	57
Indiana	Pike	No. V	Blackfoot	54
Iowa	Lucas	Cherokee	Big Ben	61
Kansas	Crawford	Bevier	Clemens	62
Kentucky	Bell	High Splint	Davisburg	44
	Muhlenburg	No. 11	Crescent	55
	Pike	Elkhorn Nos. 1 and 2	Dixie	42
Missouri	Boone	Bevier	Mark Twain	62
Montana	Richland		Savage	62
New Mexico	McKinley	Black Diamond	Sundance	51
North Dakota	Burke		Noonan	38
Ohio	Belmont	No. 9	Linda	50
	Harrison	No. 8	Bradford	51
Pennsylvania	Cambria	Lower Kittaning (bituminous)	Bird No. 2	109
	Indiana	Lower Freeport (bituminous)	Acadia	83
	Schuykill	Various (anthracite)		38
	Washington	Pittsburgh	Florence	55
	Westmoreland	Upper Freeport	Jamison	65
Tennessee	Grundy	Sewanee	Ramsey	59
Utah	Carbon	Castle Gate	Carbon	47
Virginia	Buchanan	Splash Dam	Harman	68
	Dickenson	Upper Banner		84
	Wise	Morris	Roda	43
West Virginia	Fayette	Sewell	Summerlee	86
	McDowell	Pocahontas No. 3	Jacobs Fork	96
	Wyoming	Powellton	Coal Mountain	58
	Wyoming	No. 2 Gas	Kopperston	70
Wyoming	Campbell	Smith/Rowland	Wyodak	59

Source: Baughman (1978, p. 169).

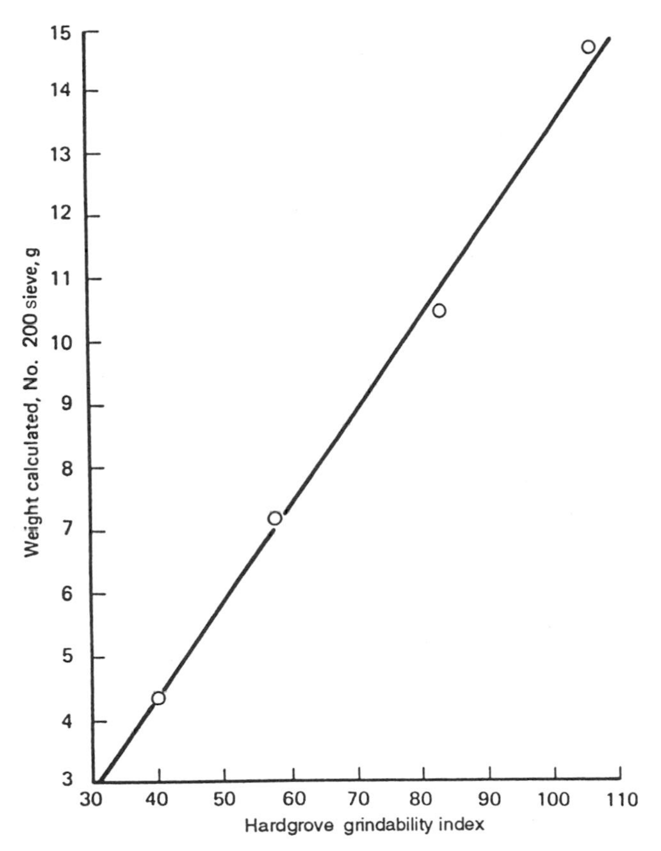

FIGURE 8.2 Calibration chart for the Hardgrove grindability index. (From ASTM, 2004.)

an original soft coal chosen as a standard coal whose grindability index was set at 100.

A general relationship exists between coal grindability and rank (Figure 8.3) insofar as the easier-to-grind coals are in the medium- and low-volatile groups; nevertheless, the relationship between grindability and rank is far too approximate to permit grindability to be estimated from coal analysis. For example, low-volatile bituminous coal usually exhibits the highest grindability index value over 100, whereas high-volatile coal ranges in grindability index from approximately 55 down to the range 35 to 40. There is, however, a correlation between friability and grindability insofar as soft, easily fractured coal generally exhibits a relatively high value for the grindability index.

Variations in the data that fall outside the experimental allowable limits may be due to factors that originate during sample preparation and handling, such as: (1) the sample moisture may not have been in equilibrium with the laboratory atmosphere; (2) the sample may have been overdried or under-air-dried; (3) excessive dust loss may have occurred during screening due to a loose-fitting

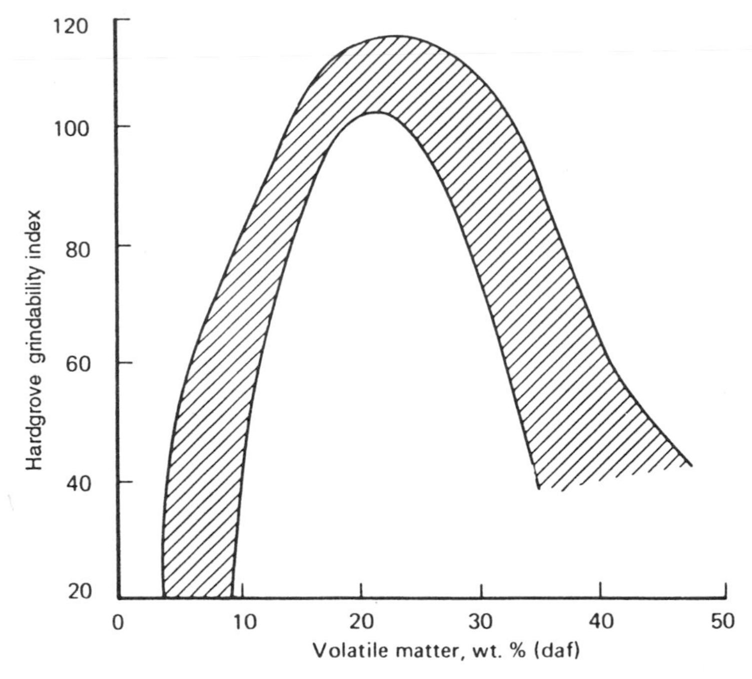

FIGURE 8.3 Variation of Hardgrove grindability index with volatile matter. (From Berkowitz, 1979, p. 91.)

pan and cover on the sieve; or (4) the sample may not have had an even distribution of particles. Moisture content is particularly troublesome in low-rank coals. The grindability index of coal varies from seam to seam and within the same seam.

8.5 DUSTINESS INDEX

Dust removal from coal preparation plants and mine workings is an important aspect of safety, and there have been constant attempts to improve dust removal technology (Henke and Stockmann, 1992). As the techniques have been developed, the predictability of how coal will behave under certain conditions, in terms of the dust produced, has also been sought.

The *dustiness index* of coal is a means of determining the relative values that represent the amount of dust produced when coal is handled in a standard manner. Thus, a 50-lb sample of coal is placed on a slide plate in a metal cabinet of prescribed size. When the plate is withdrawn, the sample falls into a drawer, and after 5 seconds, two slides are inserted into the box. The slides collect suspended dust particles for 2 minutes (coarse dust) or for 10 minutes (fine dust). The dustiness index is reported as 40 times the gram weight of dust that has settled after either 2 minutes or after an additional 8 minutes.

8.6 CLEAT STRUCTURE

A system of joint planes is often observed in coal formations, and these joint planes (cleats) are usually perpendicular to the bedding planes. Thus, cleat joints are usually vertical. The main system of joints is more commonly called the *face cleat*, whereas a cross-system of jointing is called the *butt cleat*. Furthermore, the cleat system in coal has a pronounced effect on the properties of a coal deposit. For example, holes drilled into coal perpendicular to the face cleat are said to yield from 2.5 to 10 times the amount of methane gas from the formation as holes drilled perpendicular to the butt cleat. Also, the cleat system of fracture and the frequency of cleats may determine the size of run-of-mine coal. In general, a pair of cleats will be oriented at about 90° to each other, and the orientation of the cutting elements influences the output of coal-mining machines (Figure 8.4).

Directional properties also occur in coal, and these properties can affect the direction of flow of gases. For a particular coal seam, analyses of natural fracture orientation, directional permeability, directional ultrasonic velocity, and directional tensile strength (Berenbaum and Brodie, 1959; Ingles, 1961) may disclose distinct coal structures that are important for gas flow through the coal (Skawinski et al., 1991; Puri and Seidle, 1992). It is of special importance in the passage of methane through and from coal beds, as well as in the planning of underground coal gasification tests.

8.7 DEFORMATION AND FLOW UNDER STRESS

The theology (deformation and flow) of coal has been studied in an effort to apply it to characteristics of coal in coal mines; coal elasticity (quality of regaining original shape after deformation under strain) has also been studied. However, it may be quite difficult to obtain meaningful measurements of coal elasticity. For

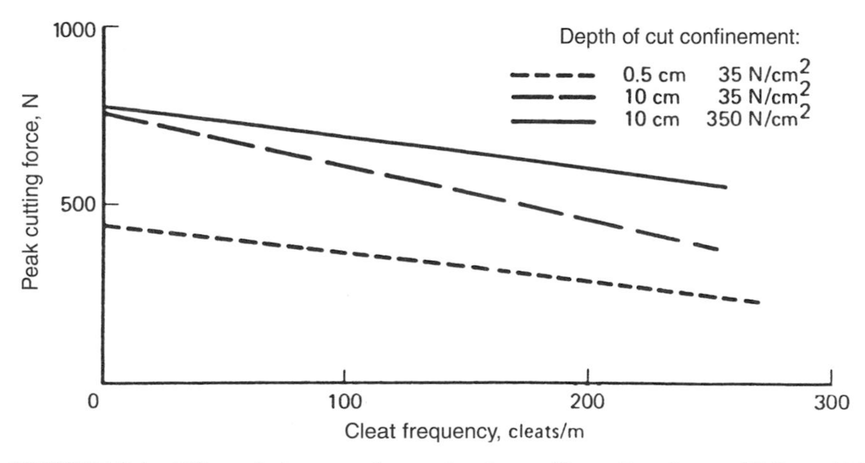

FIGURE 8.4 Effect of cleats on the cutting force. (From Baughman, 1978, p. 171.)

example, the heat that is generated in cutting a uniform sample (usually by means of a grinding wheel consisting of grit embedded in rubber) can cause plastic deformation of the coal surface and thereby affect the plastic properties. Water used to cool the grinding interface and to carry away particles may be absorbed by the coal and affect the elastic properties. Furthermore, discontinuities in the coal structure give a wide sample-to-sample variation.

REFERENCES

ASTM. 2004. *Annual Book of ASTM Standards*, Vol. 05.06. American Society for Testing and Materials, West Conshohocken, PA. Specifically:

ASTM D-409. Standard Test Method for Grindability of Coal by the Hardgrove-Machine Method.

ASTM D-440. Standard Test Method of Drop Shatter Test for Coal.

ASTM D-441. Standard Test Method of Tumbler Test for Coal.

ASTM D-3038. Standard Test Method for Drop Shatter Test for Coke.

ASTM D-3402. Standard Test Method for Tumbler Test for Coke.

ASTM D-5003. Standard Test Method for the Hardgrove Grindability Index (HGI) of Petroleum Coke.

ASTM D-5341. Standard Test Method for Measuring Coke Reactivity Index (CRI) and Coke Strength After Reaction (CSR).

Baughman, G. L. 1978. *Synthetic Fuels Data Handbook*. Cameron Engineers, Denver, CO.

Berenbaum, R., and Brodie, J. 1959. *Br. J. Appl. Phys.*, 10:281.

Berkowitz, N. 1979. *An Introduction to Coal Technology*. Academic Press, San Diego, CA.

Brown, R. L., and Hiorns, F. J. 1963. In *Chemistry of Coal Utilization*, Suppl. Vol., H. H. Lowry (Editor). Wiley, Hoboken, NJ, Chap. 3.

Chakravorty, R. N. 1984. Report ERP/CRL 84-14. Coal Research Laboratories, Canada Center for Mineral and Energy Technology, Ottawa, Ontario, Canada.

Chakravorty, R. N., and Karr, K. 1986. Report ERP/CRL 86–151. Coal Research Laboratories, Canada Center for Mineral and Energy Technology, Ottawa, Ontario, Canada.

Edwards, G. R., Evans, T. M., Robertson, S. D., and Summers, C. W. 1980. *Fuel*, 59:826.

Evans, D. G., and Allardice, D. J. 1978. In *Analytical Methods for Coal and Coal Products*, Vol. I, C. Karr, Jr. (Editor). Academic Press, San Diego, CA, Chap. 3.

Hardgrove, R. M. 1932. *Trans. Am. Soc. Mech. Eng.*, 54:37.

Henke, B., and Stockmann, H. W. 1992. *Glueckauf-Forschungsh*, 53(1):249.

Ingles, O. G. 1961. In *Agglomeration*, W. A. Knepper (Editor). Wiley, Hoboken, NJ, p. 29.

ISO. 2003. *Standard Test Methods for Coal Analysis*. International Organization for Standardization, Geneva, Switzerland. Specifically:

ISO 5074. Determination of the Hardgrove Grindability Index.

Jones, J. C., and Vais, M. 1991. *J. Hazard. Mater.*, 26:203.

Puri, R., and Seidle, J. P. 1992. *In Situ*, 16:183.

Skawinski, R., Zolcinska, J., and Dyrga, L. 1991. *Arch. Mineral. Sci.*, 36:227.

Trollope, D. H., Rosengren, K. J., and Brown, E. T. 1965. *Geotechnique*, 57:363.

van Krevelen, D. W. 1957. *Coal: Aspects of Coal Constitution*. Elsevier, Amsterdam.

Yancey, H. R., and Geer, M. R. 1945. In *Chemistry of Coal Utilization*, H. H. Lowry (Editor). Wiley, Hoboken, NJ, p. 145.

9 Spectroscopic Properties

Coal is an organic rock that is composed of macerals and minerals (Given, 1988; Speight, 1994a, and references cited therein). The precursors to coal are as diverse as the plant chemicals themselves, with the added recognition that there may have been some evolution of these chemicals over geological time as plants and their constituents have evolved to modern-day counterparts. Thus, coal is a complex chemical material wherein the constituents of the original chemical precursors have undergone considerable change through chemical and physical interactions with their environment.

Identification of the constituents of complex materials such as coal may proceed in a variety of ways but generally can be classified into three methods: (1) spectroscopic techniques, (2) chemical techniques, and (3) physical property methods whereby various structural parameters are derived from a particular property by a sequence of mathematical manipulations. It is difficult to completely separate these three methods of structural elucidation, and there must, by virtue of need and relationship, be some overlap. Thus, although this review is more concerned with the use of spectroscopic methods applied to the issues of coal structure, there will also be reference to the other two related methods.

It is the purpose of this chapter to present some indication of the spectroscopic methods that have been applied to coal analysis (Vorres, 1993, and references cited therein under the specific spectroscopic method). However, the standard test methods are not in any great abundance, but that does not stop the researcher for making such a request of the analyst for spectroscopic analysis. Thus, reference to the scientific literature is necessary, keeping in mind that the focus of the references is the description of the technique and that sample preparation and sampling are of the utmost importance.

9.1 INFRARED SPECTROSCOPY

Of all the physical techniques, infrared (IR) spectroscopy gives the most valuable information about the constitution of organic materials. Indeed, qualitative information about specific structural and functional elements can often be deduced even though the spectra are too complex for individual compound analysis. With regard to quantitative evaluation of the constituents of coal by infrared

Handbook of Coal Analysis, by James G. Speight
ISBN 0-471-52273-2

spectroscopy, the presence of mineral matter offers a constraint insofar as distortion of the peaks arising from the organic constituents is inevitable.

In addition, one of the issues as it relates to coal science is the absence of specific standard test methods that can be applied to the investigation of coal properties by infrared spectroscopy as well as by other spectroscopic methods. There are, however, test methods that are applicable to the infrared analytical technique that should be followed when the method is applied to coal analysis.

One available test method (ASTM D-1655), in which are described the standard practices for infrared multivariate quantitative analysis for turbine fuels, does allow the potential for presentation to other fuels. These practices cover a guide for the multivariate calibration of infrared spectrometers used in determining the physical or chemical characteristics of materials. These practices are applicable to analyses conducted in the near-infrared (NIR) spectral region (roughly 780 to 2500 nm) through the mid-infrared (MIR) spectral region (roughly 4000 to 400 cm^{-1}). Although the practices described deal specifically with mid- and near-infrared analysis, much of the mathematical and procedural detail contained herein is also applicable for multivariate quantitative analysis done using other forms of spectroscopy. These practices cover techniques that are applied routinely in the near- and mid-infrared spectral regions for quantitative analysis. The practices outlined cover the general cases for coarse solids and for fine-ground solids but do require the use of a computer for data collection and analysis. Considering the complexity of coal, a computerized data collection system is a benefit. This method should be used in conjunction with a test method that relates to the techniques most often used in infrared quantitative analysis and are associated with the collection and analysis of data on a computer as well as practices that do not use a computer (ASTM E-168).

In addition, the test method that covers the standard practice for general techniques of obtaining infrared spectra for qualitative analysis (ASTM E-1252) is also valuable. This test method covers the spectral range from 4000 to 50 cm^{-1} and includes techniques that are useful for qualitative analysis of liquid-, solid-, and vapor-phase samples by infrared spectrometric techniques for which the amount of sample available for analysis is not a limiting factor. These techniques are often also useful for recording spectra at frequencies higher than 4000 cm^{-1} in the near-infrared region. A separate method (ASTM E-334) is available for use in analyzing microgram quantities of samples by infrared spectrophotometric techniques.

In all cases, specific test methods (ASTM E-168; ASTM E-573; ASTM E-1252) should be consulted for theoretical aspects of the methodology, for general techniques of sample preparation, and for data workup and calculations. From the results of infrared spectroscopic investigations, it has been established (Speight, 1978, 1994b) that coal contains various aliphatic and aromatic carbon–carbon and carbon–hydrogen functions but few, if any, isolated olefin bonds (>C=C<) and acetylene (–C≡C–) bonds. The assignment of absorptions in the infrared spectra to various oxygen functions has also received some attention (Speight, 1978, 1994b).

It has also been reported that the ratio of aromatic hydrogen to total hydrogen increases with increasing rank and that at a 94% carbon content coal is completely aromatic. It has also been suggested that the percentage of the hydrogen contained in methyl groups probably lies in the range 15 to 25% and that the methyl content decreases with increasing rank of the coal. Other observations on the aromaticity of coal come from investigations of a variety of coal fractions (in particular, the optical densities of the two peaks at about 3030 and 2920 cm^{-1}).

Infrared absorption is one of three standard test methods for sulfur in the analysis sample of coal and coke using high-temperature tube furnace combustion methods (ASTM D-4239). Determination of sulfur is, by definition, part of the ultimate analysis of coal (Chapter 4), but sulfur analysis by the infrared method is also used to serve a number of interests: evaluation of coal preparation, evaluation of potential sulfur emissions from coal combustion or conversion processes, and evaluation of the coal quality in relation to contract specifications, as well as other scientific purposes. Infrared analysis provides a reliable, rapid method for determining the concentration of sulfur in coal and is especially applicable when results must be obtained rapidly for the successful completion of industrial, beneficiation, trade, or other evaluations.

In the method, the sample is burned in a tube furnace at a minimum operating temperature of 1350°C (2462°F) in a stream of oxygen to oxidize the sulfur. Moisture and particulates are removed from the gas by traps filled with anhydrous magnesium perchlorate. The gas stream is passed through a cell in which sulfur dioxide is measured by an infrared absorption detector. Sulfur dioxide absorbs infrared energy at a precise wavelength within the infrared spectrum. Energy is absorbed as the gas passes through the cell body in which the infrared energy is being transmitted: thus, less energy is received at the detector. All other infrared energy is eliminated from reaching the detector by a precise wavelength filter. Thus, the absorption of infrared energy can be attributed only to sulfur dioxide, whose concentration is proportional to the change in energy at the detector. One cell is used as both a reference and a measurement chamber. Total sulfur as sulfur dioxide is detected on a continuous basis. This method is empirical; therefore, the apparatus must be calibrated by the use of certified reference materials.

Infrared spectroscopy is also used (ASTM D-5016) to determine sulfur in coal ash in conjunction with high-temperature tube furnace combustion. This test method also allows for rapid determination of sulfur in the ash and may be used as an alternative test method (cf. ASTM D-1757). In the test, a weighed test portion is mixed with a promoting agent (which aids in the quantitative release of all sulfur present in the test portion as sulfur dioxide) and ignited in a tube furnace at a minimum operating temperature of 1350°C (2462°F) in a stream of oxygen. The combustible sulfur contained in the test portion is oxidized to gaseous oxides of sulfur. Traps filled with anhydrous magnesium perchlorate remove moisture and particulate matter. The gas stream is passed through a cell in which sulfur dioxide is measured by an infrared absorption detector. Sulfur dioxide absorbs infrared energy at a precise wavelength within the infrared spectrum. Energy is absorbed as the gas passes through the cell body in which the infrared energy is being transmitted; thus less energy is received at the detector. All other infrared energy is eliminated from reaching the detector by a precise wavelength filter. The absorption of infrared energy can be attributed only to sulfur dioxide, whose concentration is proportional to the change in energy at the detector. One cell is used as both a reference and a measurement chamber. Total sulfur as sulfur dioxide is detected on a continuous basis.

The inception of Fourier transform methods to elucidating the nature of coal overcame the barriers that arose from many of the weaker absorption bands (Dyrcasz et al., 1984; Snyder et al., 1983; Gaines, 1988). This allowed the collection of better defined spectra by careful repetition and permits further information to be obtained about the nature of the carbon–hydrogen bonding systems and the nature of the aromatic groups as well as information about the various functional groups in coal. For example, it is possible to differentiate between the oxygen species, the aliphatic species, and the substitution patterns on aromatic ring systems.

There have been many studies using the Fourier transform method, with emphasis on changes that occur during coalification, weathering, and conversion. Major findings include identifying the presence of two- and three-ring condensed aromatic systems possibly linked through quinoid and furan bridges (Sharma and Sarkar, 1983). There has also been the interesting revelation of the relative lengths of alkyl chains in different coals (shorter in coking coal compared to noncoking coal) (Erbatur et al., 1986).

9.2 NUCLEAR MAGNETIC RESONANCE

Magnetic resonance spectroscopy has a considerable history of being applied to the issue of coal structure. However, as a historical beginning, the structural types in coal were first determined by means of statistical structural analysis (Francis, 1961). One of the first methods to supersede the statistical methods was based on proton (^{1}H) magnetic resonance, which provided a quantitative distribution of the hydrogen types in coal (Brown and Ladner, 1960; Bartle, 1988; Maciel et al., 1993).

In an early publication on the subject it was reported that in a high-rank coal, 33% of the hydrogen atoms occurred in methylene groups of bridge and alicyclic structures, whereas in a low-rank coal 67% of the hydrogen occurred in this form and/or in long chains. The nuclear magnetic resonance spectra of the products obtained from the vacuum carbonization of coal have also been studied, and it was concluded that nonaromatic hydrogen occurred almost exclusively on saturated carbon atoms as aliphatic, alicyclic, or hydroaromatic structures.

Nuclear magnetic resonance spectroscopy has proved to be of great value in fossil fuel research because it allows rapid and nondestructive determination of the total hydrogen content and distribution of hydrogen among the chemical functional groups present (Bartle and Jones, 1978; Retcofsky and Link, 1978; Petrakis and Edelheit, 1979; Snape et al., 1979; Davidson, 1980, 1986; Miknis, 1982, 1988; Calkins and Spackman, 1986; Cookson and Smith, 1987; Bartle,

1988; Kershaw, 1989; Botto and Sanada, 1992; Meiler and Meusinger, 1992). However, coal is a structurally diverse material, and caution is to be exercised in the definition of chemical shift expectations. Thus, if structural definition is to be successful, the chemical shift relationships applied to coal and to coal-derived products should be lacking in ambiguity. Sample complexity will usually introduce a variety of ambiguities.

Like infrared spectroscopy, specific test methods for recording the nuclear magnetic resonance spectroscopy of coal do not exist. It is necessary, therefore, to adapt other methods to the task at hand, provided that the necessary sample preparation protocols and instrumental protocols for recording magnetic resonance spectra are followed to the letter as proposed and described for infrared spectroscopy (Section 9.1).

One particular test method (ASTM D-5292) is designed to determine the aromatic carbon contents of hydrocarbon oils by high-resolution nuclear magnetic resonance spectroscopy using high-resolution nuclear magnetic resonance, provided that the sample is completely soluble in chloroform and carbon tetrachloride at ambient temperature. Although not specifically applicable to coal per se, the method can be used for coal extracts that are soluble in the aforesaid solvents. For pulse Fourier transform (FT) spectrometers, the detection limit is typically 0.1 mol % aromatic hydrogen atoms and 0.5 mol % aromatic carbon atoms. For continuous-wave (CW) spectrometers, which are suitable for measuring aromatic hydrogen contents only, the detection limit is considerably higher and typically 0.5 mol % aromatic hydrogen atoms. This test method is not applicable to samples containing more than 1 mass % olefin or phenol compounds. However, this test method does not cover determination of the percentage mass of aromatic compounds in oils since NMR signals from both saturated hydrocarbons and aliphatic substituents on aromatic ring compounds appear in the same chemical shift region. For the determination of mass or volume percent aromatics in hydrocarbon oils, chromatographic or mass spectrometry methods can be used.

Another method (ASTM D-4808) covers the determination of the hydrogen content of petroleum products, including vacuum residua, using a continuous-wave, low-resolution nuclear magnetic resonance spectrometer. Again, sample solubility is a criterion that will not apply to coal but will apply to coal extracts. More recent work has shown that proton magnetic resonance can be applied to solid samples and has opened a new era in coal analysis by this technique (de la Rosa et al., 1993; Jurkiewicz et al., 1993).

Along similar lines, carbon-13 magnetic resonance has been applied to coal and to the elucidation of the character of soluble coal fractions and other coal-derived materials (Bartuska et al., 1978; Ladner and Snape, 1978; Miknis, 1988). In fact, it is the advances in carbon-13 magnetic resonance that have brought new thoughts to the chemical nature of, and structural types in, coal. The technique brought with it the innovative solid-state techniques that allowed coal to be viewed in its natural solid state without invoking the criterion of sample solubility. This removed the need for dissolution of coal fractions and the often tongue-in-cheek deductions about the relationship of the soluble part of coal to the majority of the solid matrix.

The data obtained, used in conjunction with those obtained from proton magnetic resonance and from elemental analysis, indicated a typical molecule of the neutral oil to be 70% aromatic and to consist of a naphthalene ring system bearing two or three saturated side chains, each having less than three atoms. Data from the investigation of the supercritical gas extraction of coal point to a similar conclusion insofar as the structural entities can be envisaged as consisting of small aromatic ring systems, with the inclusion of hydroaromatic ring systems, and the attendant alkyl groups (predominantly methyl) as well as a variety of oxygen functions.

^{13}C magnetic resonance has been applied to coal and to the elucidation of structures of soluble coal fractions (Alemany et al., 1978; Bartle and Jones, 1978; Stock, 1988). The evidence from ^{13}C magnetic resonance studies of the extract of coal appears to favor the occurrence of small-ring systems. The mean structural unit appears to consist of two- to three-ring condensed aromatic systems with 40% of the available aromatic carbons bearing alkyl, phenolic, and/or naphthenic groups. Indeed, the mass spectrum of the extract indicated the presence of alkyl aromatic compounds having from 1 to 10 or more alkyl carbons per molecule.

A derivation of the ^{13}C magnetic resonance technique, cross-polarization ^{13}C magnetic resonance, offers several advantages over the conventional technique (Miknis et al., 1981; Smith and Smoot, 1990; Song et al., 1993). The results from this method have been used to indicate that the higher-rank coals are indeed highly aromatic substances, and cross-polarization ^{13}C magnetic resonance could be used with confidence to determine the carbon aromaticity. The technique also offers valuable information about the oxygen distribution in coal (Franco et al., 1992).

On the subject of carbon aromaticity of coal, ^{13}C magnetic resonance spectroscopy has also found use in determining the fraction of carbon atoms that are in aromatic locations (f_a) as well as attempting to define the structure of the aromatic ring system. However, there is a possibility of serious underestimation of aromaticity by this method (Miknis, 1988; Snape et al., 1989; Sfihi and Legrand, 1990).

The f_a data obtained from the cross-polarization spectra can also be used to estimate the size of the aromatic ring systems (Sfihi and Legrand, 1990). The are assumptions, however, that the nonaromatic carbons are mostly methylene ($-CH_2-$) groups and that the oxygen and about 50% of the nonaromatic carbons are bonded directly to aromatic rings. Thus, the hypothetical aromatic nuclei (H_{aru}/C_{ar}) can be estimated from the equation

$$\frac{H_{aru}}{C_{ar}} = \frac{(H - 3_{ali})(2 + O)}{C_{ar}} \tag{9.1}$$

where "ar" and "ali" denote atoms in aromatic and aliphatic groupings, respectively. The data also confirm a tendency for increased aromaticity with increasing rank of the coal.

The stable free radicals in coal can adversely affect the ^{13}C signal intensity. However, the quantitative reliability of the method can be greatly improved by prior treatment of coal with chemicals such as samarium(II) iodide (Stock et al., 1988) to selectively reduce the organic free radicals. Thus, this potential error that has influenced the quantitative aspects of the method can be reduced by the use of a novel pretreatment technique as well as an appropriate standard for measurement of chemical shifts.

^{13}C magnetic resonance studies have also been used to investigate coal in terms of structural and dynamic heterogeneity (Tekely et al., 1990; Adachi and Nakamizo, 1993; Wind et al., 1993). Coals of different rank demonstrated significant structural differences, and an increase in structural homogeneity with increase in aromaticity factor was noted. In addition, the data indicated the presence of two phases: a macromolecular/rigid component and a molecular/mobile component. In addition, electron spin resonance spectroscopy has been used to study the effects of weathering/oxidation on the mobile and network phases of coal (Khan et al., 1988). The conclusion is that oxidation influences the hydrogen-rich (mobile) phase of coal more than it does the hydrogen-poor (network) of bituminous coal.

9.3 MASS SPECTROMETRY

One of the issues that is raised when data from mass spectrometry are applied to determining the character of coal is its nonvolatility. However, this should not be a deterrent to using this valuable technique. It is the interpretation that is derived from the data that should be suspect. Many workers have erroneously assumed that the volatile material is truly representative of the nonvolatile sample, and thereby hangs the error!

The use in coal analysis of gas chromatography–mass spectroscopy as well as pyrolysis–mass spectrometry and pyrolysis gas chromatography–mass spectroscopy has enabled low-molecular-weight benzenes, phenols, and naphthalenes to be identified as well as the C_{27} and C_{29}–C_{30} hopanes and C_{15} sesquiterpenes (Gallegos, 1978; Smith and Smoot, 1990; Blanc et al., 1991).

Curie point pyrolysis mass spectrometry has also been valuable in providing information about the chemical types that are evolved during the thermal decomposition of coal (Tromp et al., 1988) and, by inference, about the nature of the potential chemical types in coal. However, absolute quantification of the product mixtures is not possible, due to the small sample size, but the composition of the pyrolysis, product mix can give valuable information about the metamorphosis of the coal precursors and on the development of the molecular structure of coal during maturation. However, as with any pyrolysis, it is very important to recognize the nature and effect that any secondary reactions have on the nature of the volatile fragments, not only individually but also collectively.

Application of mass spectrometry to the identification of methyl esters of the organic acids obtained by the controlled oxidation of bituminous coal allowed the more volatile benzene carboxylic acid esters to be identified (Studier et al., 1978). These were esters of benzene tetracarboxylic acid, terephthalic acid, toluic acid, and benzoic acid. Decarboxylation of the total acid mixture was shown to afford benzene, toluene, C_2-benzenes (i.e., ethylbenzene or xylene), C_3-benzenes, C_4-benzenes (butylbenzene), C_5-benzenes, C_7-benzenes, naphthalene, methylnaphthalene, C_2-naphthalenes, biphenyl, methylbiphenyl, C_3-biphenyl, indane, methylindane, C_2-indane, phenanthrene, and fluorene.

Tandem mass spectrometry is also developing into an important analytical method for application to coal-derived materials (Wood, 1987). The analysis of heteroatom ring species and hydrocarbon species in coal-derived liquids offers indications of the location of the heteroatoms in, or on, ring systems, as well as indications of the hydrocarbon systems.

Application of gas chromatographic/mass spectrometric analysis to acidic/basic subfractions of coal-derived asphaltenes has led to the conclusion that the asphaltenes are "made up" of one-ring and/or two-ring aromatic units that are linked by methylene chains as well as by functional groups (Koplick et al., 1984). Projection of this finding to coal itself is of interest only if it can be assumed that the internuclear bonds withstood the high temperatures and were not formed as a result of secondary and tertiary (etc.) reaction. In short, the question relates to the relationship of the structural types in the asphaltenes to those in the original coal.

9.4 ULTRAVIOLET SPECTROSCOPY

Again, in the absence of specific test methods for coal, ultraviolet spectroscopic investigations must rely on investigations applied to other substances with the criteria of sample handling and sample preparation followed assiduously. The practices to be used for recording spectra (ASTM E-169) provide general information on the techniques most often used in ultraviolet and visible quantitative analysis. The purpose is to render unnecessary the repetition of these descriptions of techniques in individual methods for quantitative analysis.

One particular test method (ASTM D-2008) covers measurement of the ultraviolet absorption of a variety of petroleum products covers, or the absorbtivity of liquids and solids, or both, at wavelengths in the region 220 to 400 nm. Use of this test method implies that the conditions of measurement (wavelength, solvent if used, sample path length, and sample concentration) are specified by reference to one of the examples of the application of this test method or by a statement of other conditions of measurement.

The ultraviolet spectra of coals, examined as suspensions in potassium bromide, show an absorption band at 2650 Å that becomes more pronounced with increasing rank of the coal. This band has been assigned to aromatic nuclei, and on the basis of data obtained from comparison between the specific extinction coefficients of coal and those of standard condensed aromatic compounds, it has been concluded that the concentration of aromatic systems in coal is lower than had previously been believed.

Other investigations have led to the conclusions that coal contains benzene and naphthalene rings, heterocyclic nitrogen, cyclic ethers, hydroxyl oxygen, and methylene groups and may even be fairly uniform in structure. In addition, the position of the maxima in the ultraviolet spectra of coal fractions appeared to indicate a mean cluster size comparable with that obtained by other methods. Thus, there was the suggestion that low-rank coals contained small aromatic nuclei of various kinds with a gradual coalescence of the units to form much larger layers of aromatic nuclei occurring as rank increases.

Ultraviolet fluorescence spectroscopy has also been employed to investigate the nature of the aromatic systems in coal and in coal-derived products (Mille et al., 1990). Examination of the pyridine extracts produced data that were characteristic of compounds having three, four, or five condensed aromatic rings. However, it was not possible to obtain an estimation of the ring size within the insoluble coal matrix.

9.5 X-RAY DIFFRACTION

X-ray scattering from coal was the subject of several early studies which led to the postulation that coal contains aromatic layers about 20 to 30 Å in diameter, aligned parallel to near-neighbors at distances of about 3.5 Å (Hirsch, 1954). Small-angle x-ray scattering, which permits characterization of the open and closed porosity of coal, has shown a wide size distribution and the radius of gyration appears to be insufficient to describe the pore size. Application of the Fourier transform technique indicated that some coals have a mesoporosity with a mean radius of 80 to 100 Å (Guet, 1990).

9.6 ELECTRON SPIN RESONANCE

Electron spin resonance was first applied to coal during the 1950s (Ingram et al., 1954; Uebersfeld et al., 1954) as a method for the determination of free-radical species in coal. Since that time, electron spin resonance has been used to compare the data for coals of different rank and to explore the potential for relating the data to the various carbon systems as well as offering valuable information about aromaticity (Toyoda et al., 1966; Retcofsky et al., 1968, 1978; Petrakis and Grandy, 1978; Kwan and Yen, 1979; Khan et al., 1988; Thomann et al., 1988; Nickel-Pepin-Donat and Rassat, 1990; Bowman, 1993; Sanada and Lynch, 1993).

Early work on a series of carbonized coals gave 3×10^{19} free radicals per gram (1 free-radical per 1600 carbon atoms). It was also established that the free-radical content of coal at first increases slowly (in the range 70 to 90% carbon) (Ladner and Wheatley, 1965), rises markedly (90 to 94% w/w C), and then decreases to limits below detectability. Thus, in a coal having 70% carbon, there is one radical per 50,000 carbon atoms, but this is increased to one radical per 1000 carbon atoms in coal with 94% w/w carbon.

Electron spin resonance *g values* that arise from spin–orbit coupling are higher than 2.0023, which is the free electron value in the absence of spin–orbit coupling. The *g* values for vitrain and fusain decrease from low to high rank with the exception of the *g* values for vitrain from very high rank coals; these are markedly higher. Organic free radicals have higher *g* values if atoms with high spin–orbit coupling constants stabilize the unpaired electron. The high *g* values for low-rank coals can be explained by localization of unpaired electrons on heteroatoms. As the carbon content increases and the oxygen content decreases, the *g* values decrease and the radicals become localized on aromatic hydrocarbons until the *g* values increase again with the formation of graphitic structures by condensation of aromatic rings (Retcofsky et al., 1979).

Electron spin resonance spectra of coals usually consist of a single line with no resolvable fine structure; however, the electron nuclear double resonance (ENDOR) technique can show hyperfine interactions not easily observable in conventional electron spin resonance spectra. Recently, this technique has been applied to coal, and it is claimed that the very observation of an ENDOR signal shows interaction between the electron and nearby protons and that the results indicate that the interacting protons are twice removed from the aromatic rings on which, it is assumed, the unpaired electron is stabilized.

REFERENCES

Adachi, Y., and Nakamizo, M. 1993 In *Magnetic Resonance of Carbonaceous Solids*, R. E. Botto and Y. Sanada (Editors). Oxford University Press, Oxford.

Alemany, L. B., King, S. R., and Stock, L. M. 1978. *Fuel*, 57:738

ASTM. 2004. *Annual Book of ASTM Standards*, Vol. 05.06. American Society for Testing and Materials, West Conshohocken, PA, Specifically:

ASTM D-1655. Standard Specification for Aviation Turbine Fuels.

ASTM D-1757. Standard Test Method for Sulfate Sulfur in Ash from Coal and Coke.

ASTM D-2008. Standard Test Method for Ultraviolet Absorbance and Absorbtivity of Petroleum Products.

ASTM D-4808. Standard Test Methods for Hydrogen Content of Light Distillates, Middle Distillates, Gas Oils, and Residua by Low-Resolution Nuclear Magnetic Resonance Spectroscopy.

ASTM D-4239. Standard Test Methods for Sulfur in the Analysis Sample of Coal and Coke Using High-Temperature Tube Furnace Combustion Methods.

ASTM D-5016. Standard Test Method for Sulfur in Ash from Coal, Coke, and Residues from Coal Combustion Using High-Temperature Tube Furnace Combustion Method with Infrared Absorption.

ASTM D-5292. Standard Test Method for Aromatic Carbon Contents of Hydrocarbon Oils by High Resolution Nuclear Magnetic Resonance Spectroscopy.

ASTM E-168. Standard Practices for General Techniques of Infrared Quantitative Analysis.

ASTM E-169. Standard Practices for General Techniques of Ultraviolet–Visible Quantitative Analysis.

ASTM E-334. Standard Practice for General Techniques of Infrared Microanalysis.

ASTM E-573. Standard Practices for Internal Reflection Spectroscopy.

ASTM E-1252. Standard Practice for General Techniques for Obtaining Infrared Spectra for Qualitative Analysis.

Attar, A. 1979. In *Analytical Methods for Coal and Coal Products*, Vol. 3, C. Karr, Jr. (Editor). Academic Press, San Diego, CA, Chap. 56.

Attar, A., and Dupuis, F. 1979. *Prepr. Div. Fuel Chem. Am. Chem. Soc.*, 24(1):166.

Attar, A., and Hendrickson, G. G. 1982. In *Coal Structure*, R. A. Meyers (Editor). Academic Press, San Diego, CA.

Barron, P. F., and Wilson, M. A. 1981. *Nature*, 289:275.

Bartle, K. D. 1988. In *New Trends in Coal Science*, Y. Yurum (Editor). Kluwer Academic, Dordrecht, The Netherlands, p. 169.

Bartle, K. D., and Jones, D. W. 1978. In *Analytical Methods for Coal and Coal Products*, Vol. 2, C. Karr, Jr. (Editor). Academic Press, San Diego, CA, Chap. 23.

Bartle, K. D., and Jones, D. W. 1983. *Trends Anal. Chem.*, 2(6):140.

Bartle, K. D., Martin. T. G., and Williams, D. F. 1975. *Fuel*, 54:226.

Bartle, K. D., Jones, D. W., and Pakdel, H. 1978. In *Analytical Methods for Coal and Coal Products*, Vol. 2, C. Karr, Jr. (Editor). Academic Press, San Diego, CA, Chap. 25.

Bartle, K. D., Ladner, W. R., Martin, T. G., Snape, C. E., and Williams, D. F. 1979. *Fuel*, 58:413.

Bartle, K. D., Jones, D. W., and Pakdel, H. 1982. In *Coal and Coal Products: Analytical Characterization Techniques*, E. L. Fuller, Jr. (Editor). Symposium Series 205. American Chemical Society, Washington, DC, Chap. 2.

Barton, W. A., and Lynch, L. J. 1989. *Energy Fuels*, 3:402.

Bartuska, V. J., Maciel, G. E., and Miknis, F. P. 1978. *Prepr. Div. Fuel Chem. Am Chem. Soc.*, 23(2):19.

Berkowitz, N. 1979. *An Introduction to Coal Technology*. Academic Press, San Diego, CA.

Berkowitz, N. 1988. In *Polynuclear Aromatic Compounds*, L. B. Ebert (Editor). Advances in Chemistry Series 217. American Chemical Society, Washington, DC, Chap. 13.

Blanc, P., Valisolalao, J., Albrecht, P., Kohut, J. P., Muller, J. F., and Duchene, J. M. 1991. *Energy Fuels*. 5:875.

Bonnett, R., Czechowski, F., and Hughes, P. S. 1991. *Chem. Geol.*, 91:193.

Botto, R. E., and Sanada, Y. (Editors). 1992. *Magnetic Resonance of Carbonaceous Solids*. Advances in Chemistry Series 229. American Chemical Society, Washington, DC.

Bowman, M. K. 1993 In *Magnetic Resonance of Carbonaceous Solids*, R. E. Botto and Y. Sanada (Editors). Oxford University Press, Oxford.

Bredenberg, J. B., Huuska, M., and Vuori, A. 1987. In *Coal Science and Chemistry*, A. Volborth (Editor). Elsevier, Amsterdam, p. 1.

Brown, J. K., and Ladner, W. R. 1960. *Fuel*, 39:87.

Burchill, P. 1987. *Proceedings of the International Conference on Coal Science*, J. A. Moulijn (Editor). Elsevier, Amsterdam, p. 5.

Calkins, W. H., and Spackman, W. 1986. *Int. J. Coal Geol.*, 6:1.

Calkins, W. H., Torres-Ordonez, R. J., Jung, B., Gorbaty, M. L., George, G. N., and Kelemen, S. R. 1992. *Energy Fuels*, 6:411.

Cantor, C. R., and Schimmel, P. R. 1980. *Biophysical Chemistry*, Parts I, II, and III. W.H. Freeman, San Francisco.

Carlson, G. A. 1992. *Energy Fuels*, 6:771.

Carlson, G. A., and Granoff, B. 1991. In *Coal Science II*, H. H. Schobert, K. D. Bartle, and L. J. Lynch (Editors). Symposium Series 461. American Chemical Society, Washington, DC, Chap. 12.

Cartz, L., and Hirsch, P. B. 1960. *Philos. Trans. R. Soc.*, A252:557.

Chaffee, A. L., and Fookes, C. J. R. 1988. *Org. Geochem.*, 12:261.

Chakrabartty, S. K., and Berkowitz, N. 1974. *Fuel*, 53:240.

Chakrabartty, S. K., and Berkowitz, N. 1976. *Fuel*, 55:362.

Ciardelli, F., and Giusti, P. (Editors). 1981. *Structural Order in Polymers*. Pergamon Press, New York.

Cookson, D. J., and Smith, B. E. 1987. In *Coal Science and Chemistry*, A. Volborth (Editor). Elsevier, Amsterdam, p. 61.

Cronauer, D. C., and Ruberto, R. G. 1977. Report EPRI-AF-442. Electric Power Research Institute, Palo Alto, CA.

Davidson, R. M. 1980. *Molecular Structure of Coal*. Report ICTIS/TRO8, International Energy Agency, London.

Davidson, R. M. 1986. *Nuclear Magnetic Resonance Studies of Coal*. Report ICTIS/TR32. International Energy Agency, London.

Davis, M. F., Quinting, G. R., Bronnimann, C. E., and Maciel, G. E. 1989. *Fuel*, 68:763.

de la Rosa, L., Pruski, M., and Gerstein, B. 1993 In *Magnetic Resonance of Carbonaceous Solids*, R. E. Botto and Y. Sanada (Editors). Oxford University Press, Oxford.

Deno, N. C., Jones, A. D., Owen, B. O., and Weinschenk, J. I. 1985. *Fuel*, 64:1286.

Derbyshire, F. J., Marzec, A., Schulten, H.-R., Wilson, M. A., Davis, A., Tekely, P., Delpeuch, J. J., Jurkiewicz, A., Bronnimann, C. E., Wind, R. A., Maciel, G. E., Narayan, R., Bartle, K. D., and Snape, C. E. 1989. *Fuel*, 68:1091.

Dyrcasz, G., Bloomquist, C., and Solomon, P. 1984. *Fuel*, 63:536.

Erbatur, G., Erbatur, O., Coban, A., Davis, M. F., and Maciel, G. E. 1986. *Fuel*, 65:1273.

Faulon, J. L., Hatcher, P. G., Carlson, G. A., and Wenzel, K. A. 1993. *Fuel Process. Technol.*, 34:277.

Fieser, L. F., and Fieser, M. 1949. *Natural Products Related to Phenanthrene*. Reinhold, New York.

Francis, W. 1961. *Coal: Its Formation and Composition*. Edward Arnold, London.

Franco, D. V., Gelan, J. M., Martens, H. J., and Vanderzande, D. J.-M. 1992. *Fuel*, 71:553.

Friedel, R. A. 1959. *J. Chem. Phys.*, 31:280.

Gaines, A. F. 1988. In *New Trends in Coal Science*, Y. Yurum (Editor). Kluwer Academic, Dordrecht, The Netherlands, p. 197.

Gallegos, E. J. 1978. In *Analytical Chemistry of Liquid Fuel Sources: Tar Sands, Oil Shale, Coal, and Petroleum*, P. C. Uden, S. Siggia, and H. B. Jensen (Editors). Advances in Chemistry Series 170. American Chemical Society, Washington, DC, Chap. 2.

Gerstein, B. C., Ryan, L. M., and Murphy, P. D. 1979. *Prepr. Div. Fuel Chem. Am. Chem. Soc.*, 24(1):90.

Gibson, J. 1978. *J. Inst. Fuel*, 51:67.

Given, P. H. 1960. *Fuel*, 39:147.

Given, P. H. 1984a. *Prog. Energy Combust. Sci.*, 10:149.

Given, P. H. 1984b. *Coal Sci.* [M. L. Gorbaty, J. W. Larsen, and I. Wender (Editors)], 3:63.

Given, P. H. 1988. In *New Trends in Coal Science*, Y. Yurum (Editor). Kluwer Academic, Dordrecht, The Netherlands, p. 1.

Given, P. H., Marzec, A., Barton, W. A., Lynch, L. J., and Gerstein, B. C. 1986. *Fuel*, 65:155.

Gorbaty, M. L., George, G. N., and Kelemen, S. R. 1991. In *Coal Science II*, H. H. Schobert, K. D. Bartle, and L. J. Lynch (Editors). Symposium Series No. 461. American Chemical Society, Washington, DC, Chap. 10.

Green, T., Kovac, J., Brenner, D., and Larsen, J. W. 1982. In *Coal Structure*, R. A. Meyers (Editor). Academic Press, San Diego, CA, p. 199.

Grimes, W. R. 1982. *Coal Sci.* [M. L. Gorbaty, J. W. Larsen, and I. Wender (Editors)], 1:21.

Grint, A., Mehani, S., Trewhella, M., and Crook, M. J. 1985. *Fuel*, 64:1355.

Guet, J. M. 1990. In *Advanced Methodologies in Coal Characterization*, H. Charcosset and B. Nickel-Pepin-Donat (Editors). Elsevier, Amsterdam, p. 103.

Gunderman, K.-D., Humke, K., Emrich, E., and Rollwage, U. 1989. *Erdoel Kohle*, 42(2):59.

Haenel, M. W. 1992. *Fuel*, 71:1211.

Hessley, R. K. 1990. In *Fuel Science and Technology Handbook*, J. G. Speight (Editor). Marcel Dekker, New York

Hill, G. R., and Lyon, L. B. 1962. *Ind. Eng. Chem.*, 54:36.

Hirsch, P. B. 1954. *Proc. Roy. Soc. (London)*, A.226:143.

Ingram, D. J. E., Tapley, J. G., Jackson, R., Bond, R. L., and Murnahgan, A. R. 1954. *Nature (London)*, 174:797.

Jurkiewicz, A., Bronnimann, C. E., and Maciel, G. E. 1993 In *Magnetic Resonance of Carbonaceous Solids*, R. E. Botto and Y. Sanada (Editors). Oxford University Press, Oxford.

Kershaw, J. R. 1989. In *Spectroscopic Analysis of Coal Liquids*. Elsevier, Amsterdam, Chap. 8.

Khan, M. R., Usmen, R., Beer, N. S., and Chisholm, W. 1988. *Fuel*, 67:1668.

Koplick, A. J., Galbraith, M. N.; Salivin, I., Vit, I., and Wailes, P. C. 1984. *Fuel*, 63:1570.

Kreulen, D. J. W. 1948. *Elements of Coal Chemistry*. Nijgh & van Ditmar, Rotterdam, The Netherlands.

Kwan, C. L., and Yen, T. F. 1979. *Anal. Chem.*, 51:1225.

Ladner, W. R., and Snape, C. E. 1978. *Fuel*, 57:658.

Ladner, W. R., and Wheatley, R. 1965. *Mon. Bull. Br. Coal Utilization Res. Assoc.*, 29(7):201.

Larsen, J. W. 1988. In *New Trends in Coal Science*, Y. Yurum (Editor). Kluwer Academic, Dordrecht, The Netherlands, p. 73.

Litke, R., Leythaeuser, D., Radke, M., and Schaefer, R. G. 1990. *Org. Geochem.*, 16:247.

Long, R. B. 1979. *Prepr. Div. Petrol. Chem. Am. Chem. Soc.*, 24(4):891.

Maciel, G. E., Bronnimann, C. E., and Ridenour, C. F. 1993. In *Magnetic Resonance of Carbonaceous Solids*, R. E. Botto and Y. Sanada (Editors). Oxford University Press, Oxford.

Mallya, N., and Zingaro, R. A. 1984. *Fuel*, 63:423.

Mazumdar, B. K., Chakrabartty, S. K., and Lahiri, A. 1962. *Fuel*, 41:129.

Meiler, W., and Meusinger, R. 1992. *Annu. Rep. NMR Spectrosc.*, 24:331.

Miknis, F. P. 1982. *Magn. Reson. Rev.*, 7(2):87.

Miknis, F. P. 1988. In *New Trends in Coal Science*, Y. Yurum (Editor). Kluwer Academic, Dordrecht, The Netherlands, p. 117.

Miknis, F. P., Sullivan, M. J., Bartuska, V. J., and Maciel, G. E. 1981. *Org. Geochem.*, 3(1):19.

Mille, G., Kister, J., Doumenq, P., and Aune, J. P. 1990. In *Advanced Methodologies in Coal Characterization*, H. Charcosset and B. Nickel-Pepin-Donat (Editors). Elsevier, Amsterdam, p. 235.

Nickel-Pepin-Donat, B., and Rassat, A. J. A. 1990. In *Advanced Methodologies in Coal Characterization*, H. Charcosset and B. Nickel-Pepin-Donat (Editors). Elsevier, Amsterdam, p. 149.

Nishioka, M. 1992. *Fuel*, 71:941.

Olcay, A. 1988. In *New Trends in Coal Science*, Y. Yurum (Editor). Kluwer Academic, Dordrecht, The Netherlands, p. 401.

Painter, P. C., Coleman, M. M., Snyder, R. W., Mahajan, O. P., Komatsu, M., and Walker, P. L., Jr. 1981. *Appl. Spectrosc.*, 35(1):106.

Peppas, N. A., and Lucht, L. M. 1984. *Chem. Eng. Commun.*, 30:291.

Petrakis, L., and Edelheit, E. 1979. *Appl. Spectrosc. Rev.*, 15(2):195.

Petrakis, L., and Grandy, D. W. 1978. *Anal. Chem.*, 50:303.

Pickel, W., and Gotz, G. K. E. 1991. *Org. Geochem.*, 17:695.

Pitt, G. J. 1979. In *Coal and Modern Coal Processing: An Introduction*, G. J. Pitt and G. R. Millward (Editors). Academic Press, San Diego, CA, Chap. 2.

Redlich, P., Jackson, W. R., and Larkins, F. P. 1985. *Fuel*, 64:1383.

Retcofsky, H. L., and Friedel, R. A. 1973. *J. Phys. Chem.*, 77:68.

Retcofsky, H. L., and Link, T. A. 1978. In *Analytical Methods for Coal and Coal Products*, Vol. II, C. Karr, Jr. (Editor). Academic Press, San Diego, CA, Chap. 24.

Retcofsky, H. L., Stark, M. J., and Friedel, R. A. 1968. *Anal. Chem.*, 40:1699.

Retcofsky, H. L., Thompson, G. P., Hough, M., and Friedel, R. A. 1978. In *Organic Chemistry of Coal*, J. W. Larsen (Editor). Symposium Series 71. American Chemical Society, Washington, DC, p. 142.

Retcofsky, H. L., Hough, M., and Clarkson, R. B. 1979. *Prepr. Div. Fuel Chem. Am. Chem. Soc.*, 24(1):83.

Rouzaud, J. N., and Oberlin, A. 1990. In *Advanced Methodologies in Coal Characterization*, H. Charcosset and B. Nickel-Pepin-Donat (Editors). Elsevier, Amsterdam, p. 311.

Sanada, Y., and Lynch, L. J. 1993 In *Magnetic Resonance of Carbonaceous Solids*, R. E. Botto and Y. Sanada (Editors). Oxford University Press, Oxford.

Schafer, H. N. S. 1970. *Fuel*, 49:197.

Scott, A. C. (Editor) 1987. *Coal and Coal-Bearing Strata: Recent Advances*. Blackwell Scientific, Oxford.

Sfihi, H., and Legrand, A. P. 1990. In *Advanced Methodologies in Coal Characterization*, H. Charcosset and B. Nickel-Pepin-Donat (Editors). Elsevier, Amsterdam, p. 115.

Sharma, D. K. 1988. *J. Indian Chem. Soc.*, 65:582.

Sharma, D. K., and Sarkar, M. K. 1983. *Indian J. Technol.*, 21:24.

Sinninghe Damste, J. S., and de Leeuw, J. W. 1992. *Fuel Process. Technol.*, 30:109.

Siskin, M., and Aczel, T. 1983. *Fuel*, 62:1321.

Smith, K. L., and Smoot, L. D. 1990. *Prog. Energy Combust. Sci.*, 16:1.

Snape, C. E. 1987. *Fuel Process. Technol.*, 15:257.

Snape, C. E., Ladner, W. R., and Bartle, K. D. 1979. *Anal. Chem.*, 51:2189.

Snape, C. E., Axelson, D. E., Botto, R. E., Delpeuch, J. J., Tekely, P., Gerstein, B. C., Pruski, M., Maciel, G. E., and Wilson, M. A. 1989. *Fuel*, 68:547.

Snyder, R. W., Painter, P. C., Havens, J. R., and Koenig, G. 1983. *Appl. Spectrosc.*, 37:497.

Solomon, P. R. 1981. In *New Approaches in Coal Chemistry*, B. D. Blaustein, B. C. Bockrath, and S. Friedman (Editors). Symposium Series 169. American Chemical Society, Washington, DC, p. 61.

Solomon, P. R., Best, P. E., Yu, Z. Z., and Charpenay, S. 1992. *Energy Fuels*, 6:143.

Song, C., Hou, L., Saini, A. K., Hatcher, P. G., and Schobert, H. H. 1993. *Fuel Process. Technol.*, 34:249.

Speight, J. G. 1971. *Appl. Spectrosc. Rev.*, 5:211.

Speight, J. G. 1978. In *Analytical Methods for Coal and Coal Products*, Vol. II, C. Karr, Jr. (Editor). Academic Press, San Diego, CA, Chap. 22.

Speight, J. G. 1983. *The Chemistry and Technology of Coal*, Marcel Dekker, New York.

Speight, J. G. 1987. In *Coal Science and Chemistry*, A. Volborth (Editor). Elsevier, Amsterdam, p. 183.

Speight, J. G. 1994a. *The Chemistry and Technology of Coal*, 2nd ed. Marcel Dekker, New York.

Speight, J. G. 1994b. *Appl. Spectrosc. Rev.*, 29:117.

Speight, J. G. 1999. *The Chemistry and Technology of Petroleum*, 3rd ed. Marcel Dekker, New York.

Stadelhofer, J. W., Bartle, K. D., and Matthews, R. S. 1981. *Erdoel Kohle*, 34:71.

Stock, L. M. 1988. In *New Trends in Coal Science*, Y. Yurum (Editor). Kluwer Academic, Dordrecht, The Netherlands, p. 287.

Stock, L. M., Muntean, J. V., and Botto, R. E. 1988. In *New Trends in Coal Science*, Y. Yurum (Editor). Kluwer Academic, Dordrecht, The Netherlands, p. 1.

Studier, M. H., Hyatsu, R., and Winans, R. E. 1978. In *Analytical Methods for Coal and Coal Products*, Vol. 2, C. Karr, Jr. (Editor). Academic Press, San Diego, CA, Chap. 21.

Supaluknari, S., Larkins, F. P., Redlich, P., and Jackson, W. R. 1988. *Fuel Process. Technol.*, 19:123.

Tekely, P., Nicole, D., and Delpuech, J. J. 1990. In *Advanced Methodologies in Coal Characterization*, H. Charcosset and B. Nickel-Pepin-Donat (Editors). Elsevier, Amsterdam, p. 135.

Thomann, H., Silbernagel, B., Jin, H., Gebhard, L., Tindall, P., and Dyrkacz, G. 1988. *Energy Fuels*, 2:333.

Toyoda, S., Sugawara, S., and Honda, H. 1966. *J. Fuel Soc. Jpn.*, 45:876.

Tromp, P. J. J., Moulijn, J. A., and Boon, J. J. 1988. In *New Trends in Coal Science*, Y. Yurum (Editor). Kluwer Academic, Dordrecht, The Netherlands, p. 241.

Uebersfeld, J., Etienne, A., and Combrisson, J. 1954. *Nature (London)*, 174:614.

Vahrman, M. 1970. *Fuel*, 49:5.

Volborth, A. 1979. In *Analytical Methods for Coal and Coal Products*, Vol. 3, C. Karr, Jr. (Editor). Academic Press, San Diego, CA, Chap. 55.

Vorres, K. S. 1993. *Users' Handbook for the Argonne Premium Coal Sample Program*. Argonne National Laboratory, Argonne, IL; National Technical Information Service, U.S. Department of Commerce, Springfield, VA.

Weiss, U., and Edwards, J. M. 1980. *The Biosynthesis of Aromatic Compounds*. Wiley, Hoboken, NJ.

Wender, I., Heredy, L. A., Neuworth, M. B., and Dryden, I. G. C. 1981. In *Chemistry of Coal Utilization*, 2nd Suppl. Vol., M. A. Elliott (Editor). Wiley, Hoboken, NJ, Chap. 8.

Whitehurst, D. D., Mitchell, T. O., and Farcasiu, M. 1980. *Coal Liquefaction: The Chemistry and Technology of Thermal Processes*. Academic Press, San Diego, CA.

Winans, R. E., Melnikov, P. E., and McBeth, R. L. 1992. *Prepr. Div. Fuel Chem. Am. Chem. Soc.*, 37(2):693.

Wind, R. A., Maciel, G. E., and Botto, R. E. 1993 In *Magnetic Resonance of Carbonaceous Solids*, R. E. Botto and Y. Sanada (Editors). Oxford University Press, Oxford.

Wood, K. V. 1987. In *Coal Science and Chemistry*, A. Volborth (Editor). Elsevier, Amsterdam, p. 183.

Yoshida, T., Tokuhasho, K., Narita, H., Hasegawa, Y., and Maekawa, Y. 1984. *Fuel*, 63:282.

Youtcheff, J. S., and Given, P. H. 1982. *Fuel*, 61:980.

Youtcheff, J. S., and Given, P. H. 1984. *Prepr. Div. Fuel Chem. Am. Chem. Soc.*, 29(5): 1.

Yun, Y., Meuzelaar, H. L. C., Simmleit, N., and Schulten, H.-R. 1991. In *Coal Science II*, H. H. Schobert, K. D. Bartle, and L. J. Lynch (Editors). Symposium Series 461. American Chemical Society, Washington, DC, Chap. 8.

10 Solvent Properties

Solvent extraction is accomplished by contacting coal with a solvent and separating the residual coal material from the solvent and the extracts (Vorres, 1993, and references cited therein). However, extraction is, typically, mass transfer limited, so thorough mixing of the solvent and coal is required. Briefly, the extraction solvent is well mixed with the coal to allow potentially soluble constituents to transfer to the solvent. The residual coal and solvent are then separated by physical methods, such as gravity decanting, filtering, or centrifuging. Distillation regenerates the solvent from the material extracted.

Nondestructive solvent extraction of coal is the extraction of soluble constituents from coal under conditions where thermal decomposition does not occur. On the other hand, solvolysis (destructive solvent extraction) refers to the action of solvents on coal at temperatures at which the coal substance decomposes and in practice relates in particular to extraction at temperatures between 300 and 400°C (572 and 752°F). In the present context (i.e., the solvents extraction of coal), the solvent power of the extracting liquid appears to be determined solely by the ability of the solvent to alter the coal physically (by swelling). In this respect, the most effective solvents are aromatics, phenol derivatives, naphthol derivatives, anthracene, and phenanthrene.

For the purposes of this chapter, the solvent extraction of coal is limited to those investigations and test methods that are separate form the high-temperature treatment of coal with solvents in which the production of liquid products (liquefaction) is the goal.

The solvent extraction of coal has been employed for many years (Fremy, 1861; de Marsilly, 1862) as a means of studying the constitution of coal, with the yield and the nature of the extract dependent on the solvent type, extraction conditions, and, last but not least, the coal type (Bedson, 1902; Kreulen, 1948; van Krevelen, 1957; Francis, 1961; Dryden, 1963; Wise, 1971; Hombach, 1980; Pullen, 1983; Sternberg et al., 1983; Given, 1984; Litke et al., 1990; Blanc et al., 1991; Pickel and Gotz, 1991). The solvent extraction of coal should not be confused with the principles that lie behind coal cleaning or coal washability (Chapter 2) even though solvents (more frequently referred to as *heavy liquids*) may be used. The approach uses a different perspective and the outcome is usually quite different.

Handbook of Coal Analysis, by James G. Speight
ISBN 0-471-52273-2

There have been many attempts to define solvent behavior in terms of one or more physical properties of the solvent, and not without some degree of success. However, it is essential to note that the properties of the coal also play an important role in defining behavior of a solvent, and it has been reported that the relative solvent powers of two solvents may be reversed from one coal type to another. Thus, two properties that have found some relevance in defining solvent behavior with coal (as well as with other complex carbonaceous materials, such as petroleum asphaltenes) are the surface tension and the *internal pressure* (Speight, 1994, p. 201). However, the solvent power of primary aliphatic amines (and similar compounds) for the lower-rank coals has been attributed to the presence of an unshared pair of electrons (on the nitrogen atom).

Early work on the solvent extraction of coal was focused on an attempt to separate from coal a *coking principle* (i.e., the constituents believed to be responsible for coking and/or caking properties). But solvent extraction has actually been used to demonstrate the presence in coal of material that either differed from the bulk of the coal substance or was presumed to be similar to the bulk material.

An example of the difference of the solvent extracts from the bulk material comes from a series of studies on the exhaustive extraction of coal by boiling pyridine and fractionation of the regenerated soluble solids by sequential selective extraction schemes (Berkowitz, 1979). Subsequent analyses showed that the petroleum ether-soluble material was mostly composed of hydrocarbons (e.g., paraffins, naphthenes, and terpenes), while the ether-soluble, acetone-soluble, and acetone-insoluble fractions were resinlike substances with 80 to 89% carbon and 8 to 10% hydrogen. Indeed, this and later work (Vahrman, 1970) led to the concept that coal is a two-component or two-phase system (Derbyshire et al., 1991; Yun et al., 1991).

10.1 TEST METHODS

There are no specific test methods that apply to the solvent extraction of coal. However, procedures and protocols can be gleaned from other methods, provided that sampling and handling of the coal are in accordance with accepted standard test methods. There are several test methods that advocate the use of solvents for various reasons, and they are worthy of mention here insofar as each of the methods can be modified for application to coal. One such method (ASTM D-5369) describes standard procedures for extracting nonvolatile and semivolatile organic compounds from solids such as soil, sediment, sludge, and granular waste using Soxhlet extraction. The sample must be suitable to be mixed with the sample drying agent, sodium sulfate or magnesium sulfate, to provide drying of all sample surfaces. This practice, when used in conjunction with another test method (ASTM D-5368), is applicable to the determination of the total solvent extractable content (TSEC) of a soil, sediment, sludge, or granular solid waste but does depend on the solvent chosen. However, the practice is limited to solvents having boiling points below the boiling point of water at ambient pressure. But the method does provide sample extracts suitable for analysis by various techniques, such as gas chromatography with flame ionization detection (GC/FID) or gas chromatography with mass spectrometric detection (GC/MS).

A related test method (ASTM D-5368) describes the standard procedures for gravimetrically determining the total nonvolatile and semivolatile organic content of solvent extracts from soils or solid wastes. As written, the test method is used after a solvent extract is obtained from a soil or solid waste. For these methods to be applicable, the extraction solvent must have a boiling point less than that of water at ambient pressure. Again, the total solvent extractable content (TSEC) of a soil, sediment, sludge, or solid waste depends on the solvent and method used for the extraction.

In both of the foregoing cases, there may be some hesitancy because of the restrictions of the boiling point of the solvent. However, it is the method that should be considered and modified and applied accordingly to coal.

A test method for the batch extraction of treated or untreated solid waste or sludge, or solidified waste, to provide an indication of the leaching potential (ASTM D-5233) is also available. The goal of this test method is to provide an extract for measurement of the concentration of various analytes and therefore may be applied to a study of the smaller molecules that reside within the coal matrix. This test method, as written, is intended to provide an extract suitable for measurement of the concentration of analytes that will not volatilize under the conditions of the test method and may appear to offer limitations on the use of coal, but the test method does describe a procedure for performing a batch extraction of a solid. Again, the sampling and handling requirements that may be associated with the analysis of coal should also be applied to the method.

Once a suitable method of coal exaction has been selected, the common equipment for the method is a Soxhlet extractor, sometime fitted with a Dean and Stark adaptor. The latter allows the removal of water as a separate phase. Beyond this, and the smoking and handling constraints (remembering that coal oxidation will influence its extractability), a solvent that is suitable to the task (as defined by the objective) should be selected. In general terms, solvents for coal extraction can be grouped into four general categories: (1) nonspecific, (2) specific, (3) degrading, and (4) reactive (Williams et al., 1987). Such solvents are used in a variety of solvent-enhanced liquefaction operations. However, in terms of definition, the solvents can be described as follows:

1. Nonspecific solvents extract a small amount (10%) of coal at temperatures up to about 100°C (212°F). The extract is thought to arise from the resins and waxes that do not form a major part of the coal substance. Ethanol is an example of the nonspecific solvents.
2. Specific solvents extract 20 to 40% of the coal at temperatures below 200°C (392°F), and the nature of the extract is believed to be similar to, or even represent, the original coal. Pyridine is an example of such a solvent.
3. Degrading solvents extract major amounts of the coal (up to more than 90%) at temperatures up to 400°C (750°F). This type of solvent can be recovered substantially unchanged from the solution, and the solvent action

is presumed to depend on mild thermal degradation of the coal to produce smaller soluble fragments. Phenanthrene is an example of such a solvent.

4. Reactive solvents dissolve coal by active interaction. Such solvents are usually hydrogen donors (e.g., tetralin, 1,2,3,4-tetrahydronaphthalene) and their chemical composition is affected appreciably during the process. Again, using tetralin as the example, the solvent is converted to the aromatic counterpart (in this case, naphthalene) and the products from the coal can vary in composition, depending on the reaction severity and the ratio of the solvent to the coal. In addition, the extracts differ markedly in properties from those obtained with degrading solvents.

Recently, considerable attention has been paid to the use of compressed gases and liquids as solvents for extraction processes (Schneider et al., 1980; Dainton and Paul, 1981; Bright and McNally, 1992; Kiran and Brennecke, 1992), although the law of partial pressures indicates that when a gas is in contact with a material of low volatility, the concentration of *solute* in the gas phase should be minimal and decrease with increased pressure. Nevertheless, deviations from this law occur at temperatures near the critical temperature of the gas, and the concentration of solute in the gas may actually be enhanced as well as increased with pressure.

The technique of extracting virtually nonvolatile substances is particularly useful for materials that decompose before reaching boiling point and is therefore well suited to the extraction of the liquids formed when coal is heated to about 400°C (750°F). Thus, supercritical gas or fluid extraction affords a means of recovering the liquid products when they are first formed, avoiding undesirable secondary reactions (such as coke formation), and yields of extract up to 25 or 30% have been recorded.

Although the yields of extract may be lower than can be obtained with some liquid solvents and the use of high pressures might appear to be disadvantageous, there are nevertheless two positive features related to supercritical gas and fluid extraction of coal: (1) the extracts have lower molecular weights (about 500 compared to greater than 2000) and higher hydrogen content and may, presumably, be more readily converted to useful products; and (2) solvent removal and recovery is more efficient, that is, a pressure reduction has the ability to precipitate the extract almost completely.

10.2 ACTION OF SPECIFIC SOLVENTS

In general circumstances, unless solvolysis is involved, the more common organic solvents, such as benzene, alkylbenzenes, methanol, acetone, chloroform, and diethyl ether, dissolve little of the true coal substance and usually extract only that material that is occluded within the coal matrix.

The effect of pyridine on coal has been known since the late days of the nineteenth century, and extensive follow-up studies were carried out to determine the comparative extractability of pyridine and chloroform (Bedson, 1902;

Cockram and Wheeler, 1927, 1931; Blayden et al., 1948; Wender et al., 1981). These studies, as well as later work (Dryden, 1950, 1951; Given, 1984), showed that significant yields of extracts, often as high as 35 to 40%, can be obtained by using pyridine, certain heterocyclic bases, or primary aliphatic amines (which may, but need not, contain aromatic or hydroxyl substituents). Secondary and tertiary aliphatic amines are often much less effective insofar as more than one alkyl group on the amine appears to present steric problems that interfere with the interaction between the solvent and the coal.

In the higher-rank bituminous coals, α- and β-naphthol are both effective solvents, but for subbituminous coal, β-naphthol may produce even less extract than phenanthrene, but α-naphthol may extract as much is 83% w/w of the coal. Ring compounds, such as phenanthrene, appear to be superior for extracting bituminous coals of medium rank.

Many organic liquids have been suspected of exercising more than a solvent action on coal. An indication of chemical interaction is the observation that the total weight of products sometimes exceeds the original weight of coal, although up to about 5% may be the result of strong adsorption of solvents on the residue and extract. With mixed solvents, the potential for interaction may be increased.

Mixtures of the higher ketones and formamides have been reported to be considerably better solvents for coal (82% C) than either of the pure components. Mixtures worthy of note are acetophenone–monomethylformamide and methylcyclohexanone–dimethylformamide (equimolecular proportions). Thus, some solvent pairs may show enhanced solvent power, whereas others may behave independently in admixture.

10.3 INFLUENCE OF COAL RANK

Coal rank has a considerable influence on the nature and amount of extracts obtained by the solvent extraction of coal (Kiebler, 1945). In addition, the soluble products of the extraction, whether they be called *extracts* or (incorrectly) *bitumen*, vary according to the means by which they are obtained.

10.3.1 Benzene-Type Solvents

The results of extracting coals with benzene and benzene–ethanol mixtures have been reported (Table 10.1), but only a broad general trend seems to emerge from these observations. Thus, it appears that for coals with more than 88% carbon content for less than 25% volatile matter) the amount of extract obtainable decreases rapidly; with coals of carbon content lower than this limit, no definite trend with rank appears to be evident.

10.3.2 Nitrogen-Containing Solvents

The first systematic work on the extraction of coal using nitrogen-containing solvents (e.g., pyridine) resulted in the production of substantial amounts of

TABLE 10.1 Extraction of Coal Using Aromatic–Hydroaromatic Solvents

Yield	Solvent	Details
Increasing with decreasing rank	Naphthalene	
	Hydrogenated pitch	
	Anthracene oil, retene, pyridine	10 coals, 14.6–43.7% volatile matter, pronounced peak in 30–40% range with rapid decrease below about 20% volatile matter
	Neutral tar oil	Indication of flat maximum between 25 and 40% volatile matter, but points very scattered
	Tetralin	Yield increasing with increasing rank, then decreasing
	Phenanthrene	Coals from lignite to anthracite maximum at coal of 84.4% C

Source: Dryden (1963).

TABLE 10.2 Extraction of Coal Using Nitrogen-Containing Solvents

Yield	Solvent	Details
Increasing with decreasing rank	Pyridine	Two coals of 90.4 and 88.5% C
	Ethylenediamine	British bright bituminous and anthracite coals, 78–94% C; particularly rapid decrease in yield between 85 and 88% C
Exhibiting no correlation	Pyridine	South Wales coals, maximum extraction at 30% volatile matter

extract, which were then fractionated further to produce a series of fractions based on their solubility/insolubility in different solvents (Burgess and Wheeler, 1911).

Since that time, a general correlation between yield of extract and carbon content has been reported for ethylenediamine (Table 10.2). Various other amine solvents (e.g., monoethanolamine) show similar behavior insofar as the extract yield may decrease markedly with increase in rank for coals under 85% carbon. On the other hand, the yield of extract using solvents such as benzylamine, piperidine, and pyridine may show much less variation with rank, and the effectiveness of many solvents may decrease markedly for coal having more than 88% carbon (van Krevelen, 1965).

Thus, it is possible to deduce several preferred options in the solvent extraction of coals using the so-called "specific solvents." For example, the yield of extract usually decreases (but may, on occasion, increase) with an increase in the carbon content of the coal over the range 80 to about 87% carbon. However, use of a solvent such as pyridine may produce anomalous results, and the petrographic composition of the coal may also have an effect. For coals having more than approximately 81% carbon, the yield of extract is diminished to such an extent that only negligible yields of extracts are noted for coals having more than approximately 93% carbon.

There is a relationship between yield of the extract and the saturation sorption or imbibition of solvent that is independent of the rank coal or the particular amine solvent (Dryden, 1951). An adsorption isotherm for ethylenediamine vapor on an 82% carbon coal exhibited three main features: (l) chemisorption up to 3 to 6% adsorbed, (2) a fairly normal sorption isotherm from the completion of chemisorption up to a relative pressure of at least 0.8, and (3) a steeply rising indefinite region near saturation that corresponded to observable dissolution of the coal.

10.4 INFLUENCE OF PETROGRAPHIC COMPOSITION

It is generally recognized that vitrain is the most soluble constituent of any particular coal, whereas fusain is the least soluble. Indeed, early work on the liquefaction of coal by dissolution in a solvent showed that even at temperatures on the order of 400°C (750°F), fusain is, to all intents and purposes, insoluble. Under these aforementioned conditions, durain did show some response to the solvent but still did not match the solubility of vitrain.

10.5 ANALYSIS OF COAL EXTRACTS

There is, on occasion, and depending on the circumstances, the need to analyze the extracts that have been retrieved from coal by the use of various solvents. This assumes that sufficient material has been isolated to proceed with the analysis. The caveat is, of course, that all of the solvent used from the extraction procedure has been removed from the extracts. If not, it may be removed by the use of fractionation techniques applied to the extracts, and corrections will need to be applied to determine the correct yield of the extract fractions. To fractionate the extracts into a variety of subfractions, the nature of the coal must be taken into account as well as the nature of the chemicals extracted from the coal.

The literature abounds with reports of the fractionation of extracts from coal, more specifically to the liquids and solids produced by various coal liquefaction processes (Speight, 1994, and references cited therein). However, to accomplish the subfractionation of the coal extracts, it is preferable to apply (with any modifications noted) a suite of the standard test methods that are often applied to petroleum and its residua. However, before application of any subfractionation procedure to coal extracts, it may be necessary to *stabilize* the extracts (at a fixed temperature and pressure) to ensure that losses of potentially volatile constituents do not occur during the subfractionation procedure. Once the extracts have been stabilized, it is in order to proceed.

The common methods used to determine composition generally rely on chromatographic fractions using an adsorbent, although other methods are available, and since a discussion of these methods is available in detail elsewhere (Speight, 2001, 2002), it is not the intent to present the details here. Only a brief outline is given and the most convenient example to use relates to the methods applied to the composition of residual fuel oil (Speight, 2002, p. 217 et seq.) that might bear a relationship to the extracts obtained from coal.

The test methods of interest for the analysis of coal extracts include tests that measure chemical composition. The preeminent test methods are those that are applied to measurement of the asphaltene content. The issue with coal extracts is that the presence of asphaltene constituents in the extracts is solvent dependent. But assuming that a test is necessary to determine that whether asphaltene constituents are or are not present, several test are methods available.

The asphaltene fraction (ASTM D-893; ASTM D-2007; ASTM D-3279; ASTM D-4124; ASTM D-6560; IP 143) is the highest molecular weight and most complex fraction. In any of the methods for determination of the asphaltene content, the sample is mixed with a large excess (usually >30 volumes hydrocarbon per volume of sample) of low-boiling hydrocarbon, such as *n*-pentane or *n*-heptane. For an extremely viscous sample, a solvent such as toluene may be used prior to addition of the low-boiling hydrocarbon, but an additional amount of the hydrocarbon (usually >30 volumes hydrocarbon per volume of solvent) must be added to compensate for the presence of the solvent. After a specified time, the insoluble material (the asphaltene fraction) is separated (by filtration) and dried. The yield is reported as a percentage (% w/w) of the original sample.

In any of these tests, different hydrocarbons (such as *n*-pentane or *n*-heptane) will give different yields of the asphaltene fraction, and if the presence of the solvent is not compensated by the use of additional hydrocarbon, the yield will be erroneous. In addition, if the hydrocarbon is not present in large excess, the yields of the asphaltene fraction will vary and will be erroneous (Speight, 1999).

Another method, not specifically described as an asphaltene separation method, is designed to remove pentane- or heptane-insoluble constituents by membrane filtration (ASTM D-4055). In the method, a sample of oil is mixed with pentane in a volumetric flask, and the oil solution is filtered through a 0.8-μm membrane filter. The flask, funnel, and filter are washed with pentane to completely transfer any particulate matter onto the filter, after which the filter (with the particulate matter) is dried and weighed to give the pentane-insoluble constituents as a percent by weight of the sample. Particulate matter in the extracts can also be determined by membrane filtration (ASTM D-2276; ASTM D-5452; ASTM D-6217; IP 415).

The composition of the extracts will vary and is dependent on coal rank and the solvent employed for the extraction. Nevertheless, the composition of the extracts can be reported in the form of four or five major fractions as deduced by adsorption chromatography (Speight, 2001, 2002).

Column chromatography is generally used for compositional analyses (ASTM D-2007; ASTM D-2549). The former method (ASTM D-2007) advocates the use of adsorption on clay and clay–silica gel followed by elution of the clay with pentane to separate saturates, elution of clay with acetone–toluene to separate polar compounds, and elution of the silica gel fraction with toluene to separate aromatic compounds. The latter method (ASTM D-2549) uses adsorption on a bauxite–silica gel column. Saturates are eluted with pentane; aromatics are eluted with ether, chloroform, and ethanol.

REFERENCES

ASTM, 2004. *Annual Book of ASTM Standards*, Vol. 05.06. American Society for Testing and Materials, West Conshohocken, PA. Specifically:

ASTM D-893. Standard Test Method for Insolubles in Used Lubricating Oils.

ASTM D-2007. Standard Test Method for Characteristic Groups in Rubber Extender and Processing Oils and Other Petroleum-Derived Oils by the Clay-Gel Absorption Chromatographic Method.

ASTM D-2276. Standard Test Method for Particulate Contaminant in Aviation Fuel by Line Sampling.

ASTM D-2549. Standard Test Method for Separation of Representative Aromatics and Nonaromatics Fractions of High-Boiling Oils by Elution Chromatography.

ASTM D-3279. Standard Test Method for *n*-Heptane Insolubles.

ASTM D-4055-04. Standard Test Method for Pentane Insolubles by Membrane Filtration.

ASTM D-4124. Standard Test Methods for Separation of Asphalt into Four Fractions.

ASTM D-5233. Standard Test Method for Single Batch Extraction Method for Wastes.

ASTM D-5368. Standard Test Methods for Gravimetric Determination of Total Solvent Extractable Content (TSEC) of Solid Waste Samples.

ASTM D-5369. Standard Practice for Extraction of Solid Waste Samples for Chemical Analysis Using Soxhlet Extraction.

ASTM D-5452. Standard Test Method for Particulate Contamination in Aviation Fuels by Laboratory Filtration.

ASTM D-6217-98. Standard Test Method for Particulate Contamination in Middle Distillate Fuels by Laboratory Filtration.

ASTM D-6560. Standard Test Method for Determination of Asphaltenes (Heptane Insolubles) in Crude Petroleum and Petroleum Products.

Bedson, P. P. 1902. *J. Soc. Chem. Ind.*, 21:241.

Berkowitz, N. 1979. *An Introduction to Coal Technology*. Academic Press, San Diego, CA.

Blanc, P., Valisolalao, J., Albrecht, P., Kohut, J. P., Muller, J. F., and Duchene, J. M. 1991. *Energy Fuels*, 5:875.

Blayden, H. E., Gibson, J., and Riley, H. L. 1948. *J. Chem. Soc.*, 1693.

Bright, F. V., and McNally, M. E. P. (Editors). 1992. *Supercritical Fluid Technology: Theoretical and Applied Approaches in Analytical Chemistry*. Symposium Series 488. American Chemical Society, Washington, DC.

Burgess, M. J., and Wheeler, R. V. 1911. *J. Chem. Soc.*, 99:649.

Cockram, C., and Wheeler, R. V. 1927. *J. Chem. Soc.*, 700.

Dainton, A. D., and Paul, P. F. M. 1981. In *Energy and Chemistry*, R. Thompson (Editor). Royal Society of Chemistry, London, p. 32.

de Marsilly, C. 1862. *Ann. Chim. Phys.*, 66(3):167.

Derbyshire, F. J., Davis, A., and Lin, R. 1991. In *Coal Science*, H. H. Schobert, K. D. Bartle, and L. J. Lynch (Editors). Symposium Series 461. American Chemical Society, Washington, DC, Chap. 7.

Dryden, I. G. C. 1950. *Fuel*, 30:39.

Dryden, I. G. C. 1951. *Nature (London)*, 16:561.

Dryden, I. G. C. 1963. In *Chemistry of Coal Utilization*, Suppl. Vol. 7, H. H. Lowry (Editor). Wiley, Hoboken, NJ, Chap. 6.

Francis, W. 1961. *Coal: Its Formation and Composition*, Edward Arnold, London, p. 452 et seq.

Fremy, E. 1861. *Compt. Rend*, 52:114.

Given, P. H. 1984. In *Coal Science*, Vol. 3, M. L. Gorbaty, J. W. Larsen, and I. Wender (Editors). Academic Press, San Diego, CA, p. 63.

Hombach, H. P.. 1980. *Fuel*, 59:465.

IP 143. Asphaltenes (Heptane Insolubles).

IP 415. Particulate Content of Gas Oil.

Kiebler, M. W. 1945. In *Chemistry of Coal Utilization*, H. H. Lowry (Editor). Wiley, Hoboken, NJ, Chap. 19.

Kiran, E., and Brennecke, J. F. (Editors). 1992. *Supercritical Fluid Engineering Science: Fundamentals and Applications*. Symposium Series 514. American Chemical Society, Washington, DC.

Kreulen, D. J. W. 1948. *Elements of Coal Chemistry*. Nijgh & van Ditmar, Rotterdam, Chap. 14.

Litke, R., Leythaeuser, D., Radke, M., and Schaefer, R. G. 1990. *Org. Geochem.*, 16:247.

Pickel, W., and Gotz, G. K. E. 1991. *Org. Geochem.*, 17:695.

Pullen, J. R. 1983. In *Coal Science*, Vol. 2, M. L. Gorbaty, J. W. Larsen, and I. Wender (Editors). Academic Press, San Diego, CA, p. 173.

Schneider, G. M., Stahl, E., and Wilke, G. 1980. *Extraction with Supercritical Gases*. Verlag Chemie, Weinheim, Germany.

Speight, J. G. 1994. *The Chemistry and Technology of Coal*, 3rd ed. Marcel Dekker, New York.

Speight, J. G. 1999. *The Chemistry and Technology of Petroleum*, 3rd ed. Marcel Dekker, New York.

Speight, J. G. 2001. *Handbook of Petroleum Analysis*. Wiley, Hoboken, NJ.

Speight, J. G. 2002. *Handbook of Petroleum Product Analysis*. Wiley, Hoboken, NJ.

Sternberg, V. I., Baltisberger, R. J., Patel, K. M., Raman, K., and Wollsey, N. F. 1983. In *Coal Science*, Vol. 2, M. L. Gorbaty, J. W. Larsen, and I. Wender (Editors). Academic Press, San Diego, CA, p. 125.

Vahrman, M. 1970. *Fuel*, 49:5.

van Krevelen, D. W. 1957. *Coal Science: Aspects of Coal Constitution*. Elsevier, Amsterdam, Chap. 5.

van Krevelen, D. W. 1965. *Fuel*, 44:229.

Vorres, K. S. 1993. *Users' Handbook for the Argonne Premium Coal Sample Program.* Argonne National Laboratory, Argonne, IL; National Technical Information Service, U.S. Department of Commerce, Springfield, VA.

Wender, I., Heredy, L. A., Neuworth, M. B., and Dryden, I. G. C. 1981. In *Chemistry of Coal Utilization*, 2nd Suppl. Vol., M. A. Elliott (Editor). Wiley, Hoboken, NJ, Chap. 8.

Williams, J. M., Vanderborgh, N. E., and Walker, R. D. 1987. In *Coal Science and Chemistry*, A. Volborth (Editor). Elsevier, Amsterdam, p. 435.

Wise, W. S. 1971. *Solvent Treatment of Coal.* Mills and Boon, London.

Yun, Y., Meuzelaar, H. L. C., Simmleit, N., and Schulten, H.-R. 1991. In *Coal Science II*, H. H. Schobert, K. D. Bartle, and L. J. Lynch (Editors). Symposium Series 461. American Chemical Society, Washington, DC Chap. 8.

Glossary

See also ASTM D-121.

Agglomerating: property of softening when coal is heated to above about 400°C (752°F) in a nonoxidizing atmosphere; coal cools to a coherent mass after cooling to room temperature; formation of larger coal or ash particles by smaller particles sticking together.

Air-dried moisture: often used (incorrectly) to refer to both residual moisture and air-dry loss.

Air drying: process of partial drying of coal to bring its moisture near to equilibrium with ambient conditions in the laboratory or the room in which further reduction and division of the sample will take place (ASTM D-2013; ASTM D-2234).

Air-drying loss: loss in mass (% by weight) resulting from air drying; will vary from laboratory to laboratory, depending on ambient conditions in the laboratory (ASTM D-3302).

Alginite: coal maceral. *See also* **Maceral**.

Analysis sample: determination of the constituents of coal. *See also* **Sample**.

Anisotropic: exhibiting optical properties of different values when viewed with an optical microscope having mutually exclusive polarized light (ASTM D-5061).

Anthracene oil: heaviest distillable coal tar fraction, with distillation range 270–400°C (520–750°F), containing creosote oil, anthracene, phenanthrene, carbazole, and so on.

Anthracite (hard coal): hard, black, shiny coal very high in fixed carbon and low in volatile matter, hydrogen, and oxygen. *See also* **Rank** (ASTM D-388).

Anthraxylon: U.S. Bureau of Mines term for vitrinite viewed by transmitted light.

Apparent rank: rank designation obtained on samples other than channel samples, but otherwise conforming to procedures of coal classification (ASTM D-388).

As-analyzed moisture: used to indicate determined moisture.

As-determined basis: *see* **Reporting**.

Handbook of Coal Analysis, by James G. Speight
ISBN 0-471-52273-2

Ash: noncombustible residue remaining after complete coal combustion; not necessarily identical, in composition or quantity, with the inorganic substances (mineral matter) (ASTM D-3174).

Ash analysis: percentages of inorganic oxides present in an ash sample. Ash analyses are used for evaluation of the corrosion, slagging, and fouling potential of coal ash. The ash constituents of interest are silica (SiO_2) alumina (Al_2O_3), titania (TiO_2), ferric oxide (Fe_2O_3), lime (CaO), magnesia (MgO), potassium oxide (K_2O), sodium oxide (Na_2O), and sulfur trioxide (SO_3). An indication of ash behavior can be estimated from the relative percentages of each constituent.

Ash fusion temperatures: set of temperatures that characterize the behavior of ash as it is heated. These temperatures are determined by heating cones of ground, pressed ash in both oxidizing and reducing atmospheres.

Ash initial deformation temperature: temperature at which coal begins to fuse and become soft.

As-mined coal: *see* **Run-of-mine coal**.

As-received basis: *see* **Reporting**.

As-received moisture: moisture present in a coal sample when delivered.

As-shipped or as-produced coal: raw or prepared coal in any state or condition at which it leaves the mine property or loading facility (ASTM D-4749).

ASTM: American Society for Testing and Materials, located in West Conshohocken, Pennsylvania.

Attritus: microscopic coal constituent composed of macerated plant debris mixed intimately with mineral matter and coalified.

Auger increment: retained portion of one extraction operation of the auger (ASTM D-4916).

Banded coal: *see* **Coal**.

Bias: difference between the population mean of the test results and an accepted reference value; a systematic error as contrasted to random error, and there may be one or more systematic error components contributing to the bias (ASTM E-456).

Belt feeder (feeder breaker): crawler-mounted surge bin often equipped with a crusher or breaker and used in room-and-pillar sections positioned at the end of the section conveyor belt. It allows quick discharge of a shuttle car. It sizes the coal, and a built-in conveyor feeds it at an appropriate rate onto the conveyor belt.

Bench: surface of an excavated area at some point between the material being mined and the original surface of the ground, on which equipment can sit, move, or operate. A working road or base below a high wall, as in contour stripping for coal.

Beneficiation: *see* **Physical coal cleaning**.

Binder phase: continuous solid carbon matrix formed during the thermoplastic deformation of those coal macerals that become plastic during carbonization; formed from the thermoplastic deformation of reactive (vitrinite and liptinite) and semi-inert (semi-fusinite) coal macerals of metallurgical bituminous coals (ASTM D-5061).

Bituminous (soft) coal: relatively soft dark brown to black coal, lower in fixed carbon than anthracite but higher in volatile matter, hydrogen, and oxygen.

Boghead coal: same as cannel coal, except that algal remains can be seen under the microscope. *See also* **Coal**.

Bone coal: *see* **Coal**.

Bottom size: coal that passes through a standard sieve with the largest openings through which passes a total of less than 15% of the sample; should not be confused with particle size (ASTM D-4749).

Bright coal: U.S. Bureau of Mines term for a combination of clarain and vitrain with small amounts of fusain.

Briquette: cylindrical block composed of granulated coal (or coke) particles compressed and embedded with a binder.

BS: British Standards Institution, located in London.

Caking coal: coal that fuses together or cakes when heated.

Calorific value: heat of combustion of a unit quantity of a substance; usually expressed in British thermal units per pound (Btu/lb); can also be calories per gram (cal/g) or joules per gram (J/g), when required (ASTM D-2015; ASTM D-3286).

Calorimeter: equipment (bomb) for determining calorific value (ASTM D-3286).

Calorimeter jacket: insulating medium surrounding the calorimeter.

Cannel coal: predominately durain with lesser amounts of vitrain than in splint coal and small quantities of fusain; spores can be seen under the microscope. *See also* **Coal**.

Carbonate carbon: carbon content present in the carbonate in the mineral matter which is noncombustible (ASTM D-6316).

Carbon form: microscopically distinguishable carbonaceous textural components of coal and coke, exuding mineral carbonates; recognized by reflectance, anisotropy, and morphology; derived from the organic portion of coal and can be anisotropic or isotropic (ASTM D-5061).

Carbonization: process whereby coal is converted to coke by devolatilization.

Carbon monoxide: colorless, odorless, very toxic gas formed by incomplete combustion of carbon, as in water gas or producer gas production.

Chance error: error that has an equal probability of being positive or negative; the mean chance error from a series of observations tends toward zero as the number of observations approaches infinity.

Circular anisotropic phase: group of binder-phase anisotropic carbon textures that are distinguished by approximately circular domains (ASTM D-5061).

Clarain: macroscopic coal constituent (lithotype) known as *bright-banded coal,* composed of alternating bands of vitrain and durain.

Coal: brown to black combustible sedimentary organic rock that is composed principally of consolidated and chemically altered plant remains (ASTM D-388).

Attrital coal: ground mass or matrix of banded coal in which vitrain and, commonly, fusain layers as well, are embedded or enclosed.

Banded coal: coal that is visibly heterogeneous in composition, being composed of layers of vitrain, attrital coal, and commonly, fusain.

Boghead coal: nonbanded coal in which the liptinite (the waxy component) is predominantly alginite.

Bone coal: impure coal that contains much clay or other fine-grained detrital mineral matter.

Cannel coal: Nonbanded coal in which the liptinite is predominantly sporinite.

Fusain: coal layers composed of chips and other fragments in which the original form of plant tissue structure is preserved; commonly has fibrous texture with a very dull luster; friable and resembles charcoal; commonly concentrated in bedding layers or lenses that form planes of weakness in coal and thus is often exposed on bedding surfaces of broken coal.

Impure coal: coal having 25 wt % or more, but less than 50 wt %, of ash on the dry basis.

Lithotypes: constituents of banded coal: vitrain, fusain, clarain, durain, or attrital coal, or a specific mixture of two or more of these.

Mineralized coal: impure coal that is heavily impregnated with mineral matter, either dispersed or discretely localized along cleat joints or other fissures; pyrite and calcareous minerals are the most common (ASTM D-2796).

Nonbanded coal: consistently fine-granular coal essentially devoid of megascopic layers; may be interbedded with common banded coal or form a discrete layer at the top or bottom of the seam, or may compose the entire seam; formed from natural accumulations of finely comminuted plant detritus and commonly includes a significant amount and variety of remains of pollen grains, spores, planktonic algae, wax and resin granules, as well as other fragments of plants.

Vitrain: shiny black bands, thicker than 0.5 mm, of subbituminous and higher-rank banded coal; attributed to the coalification of relatively large fragments of wood and bark and commonly traversed by many fine cracks oriented normal to the banding.

Coal ball: carbonate concretion formed during peat formation, usually around a leaf.

Coal bed methane: methane (natural gas, CH_4) is a by-product of the coalification of plant material. There are two types: biogenic methane produced during microbial decay of the peat (also called *swamp gas*) and thermogenic methane produced by the thermal cracking of the organic molecules during the formation of higher ranks of coal. This is the gas that has caused so many fires and explosions in underground mines. It is now beginning to be extracted and sold commercially.

Coal gas: mixture of volatile products (primarily, hydrogen, methane, carbon monoxide, and nitrogen) remaining after removal of water and tar, obtained from carbonization of coal, having a heat content of 400 to 600 Btu/ft^3.

Coal gasification: production of synthetic gas from coal.

Coal seam: layer, vein, or deposit of coal; a stratigraphic part of the Earth's surface containing coal.

Coal sizes: in the coal industry, the term "5 inches to $\frac{3}{4}$ inch" means all coal pieces between 5 inches and $\frac{3}{4}$ inch at their widest point. "Plus 5 inches" means coal pieces over 5 inches in size; "$1\frac{1}{2}$ to 0" or "$-1\frac{1}{2}$" means coal pieces $1\frac{1}{2}$ inches and under.

Coal tar: condensable distillate containing light, middle, and heavy oils obtained by carbonization of coal. About 8 gal of tar is obtained from each ton of bituminous coal.

Coal washability: determination of the theoretical limits for the removal of mineral impurities from coal by beneficiation processes that rely on specific gravity separation (ASTM D-4371).

Coarse coal: coal pieces larger than $\frac{1}{2}$ mm in size. *See also* **Sample**.

Coke: porous, solid residue resulting from the incomplete combustion of coal, used primarily in the steelmaking process.

Coke breeze: smallest fraction from the screening of coke; the residue that passes through a $\frac{1}{2}$- $\frac{3}{4}$-in. screen opening.

Coke button: button-shaped piece of coke resulting from a standard laboratory test that indicates the coking or free-swelling characteristics of coal; a coke button is usually expressed in numbers from 1 to 9, representing the size of the coke as compared to a standard; the more a coal swells and cokes, the higher the number assigned to it.

Collector: reagent used in froth flotation to promote contact and adhesion between particles and air bubbles.

Combustible carbon: carbon content remaining in the solid products derived from the combustion or reaction of coal, exclusive of carbonate in any form (ASTM D-5114; ASTM D-6316).

Combustibles: value obtained by subtracting the dry weight (in percent) of the ash (ASTM D-3174) from 100%, representing the original weight of the sample analyzed (ASTM D-5114).

Concentrate: froth product recovered in coal froth flotation.

Concretion: mass of mineral matter found in rock of a composition different from its own and produced by deposition from aqueous solution in the rock.

Conditioning agents: all chemicals that enhance the performance of collectors or frothers (ASTM D-5114).

Corrected temperature rise: temperature of the calorimeter, caused by the process that occurs inside the bomb; the observed temperature change corrected for various effects; measured either in °C or °F (ASTM D-3286).

C test: standard statistical test for homogeneity of variance.

Cutinite: *see* **Maceral**.

daf: *see* **Dry, ash-free basis**.

Dense media (heavy media): liquids, solutions, or suspensions having densities greater than that of the water.

Dense-media separation: coal-cleaning method based on density separation, using a heavy-media suspension of fine particles of magnetite, sand, or clay.

Dense medium: dense slurry formed by the suspension of heavy particles in water; used to clean coal.

Depositional carbon: group of carbon forms that are formed from cracking and nucleation of gas-phase hydrocarbon molecules during coal carbonization (ASTM D-5061).

Pyrolytic carbon: anisotropic carbon form caused by the deposition of carbon parallel to an inert substrate, causing the resulting texture to appear ribbonlike (ASTM D-5061).

Sooty carbon: isotropic carbon form comprised of approximately spherical particles less than 1 μm in diameter, sometimes referred to as *combustion black* (ASTM D-5061).

Spherulitic: spherical anisotropic carbon form caused by the deposition of carbon concentrically around a nucleus; sometimes referred to as *thermal black* (ASTM D-5061).

Devolatilization: removal of vaporizable material by the action of heat.

Dewatering: removal of water from coal by mechanical equipment such as a vibrating screen, filter, or centrifuge.

Divided sample: *see* **Sample**.

Domain: region of anisotropy in a carbon form that is distinctively marked by its isochromatic boundary and cleavage (ASTM D-5061).

Dry, ash-free (daf) basis: a coal analysis basis calculated as if moisture and ash were removed. *See also* **Reporting**.

Dry basis: *see* **Reporting**.

Dry sieving: sieving of coal after the sample has been air-dried under prescribed conditions (ASTM D-4749).

Drying: removal of water from coal by thermal drying, screening, or centrifuging.

Dull coal: coal that absorbs rather than reflects light, containing mostly durain and fusain lithotypes.

Durain: macroscopic coal constituent (lithotype) that is hard and dull gray in color.

Easily oxidized coal: low-rank coal such as subbituminous coal or lignite coals (ASTM D-3302).

Energy equivalent, heat capacity, or water equivalent: energy required to raise the temperature of the calorimeter an arbitrary unit; when multiplied by the corrected temperature rise, adjusted for extraneous heat effects, and divided by the weight of the sample, gives the gross calorific value.

Equilibrium: condition reached in air drying in which change in weight of the sample, under conditions of ambient temperature and humidity, is no more than 0.1% per hour or 0.05% per half-hour (ASTM D-2015; ASTM D-3302; ASTM D-3286).

Equilibrium moisture: moisture capacity of coal at 30°C (86°F) in an atmosphere of 95% relative humidity.

Error: difference of an observation from the best estimate obtainable of the true value (ASTM D-2234; ASTM D-4916).

Excess moisture: surface moisture.

Exinite: microscopic coal constituent (maceral) or maceral group containing spores and cuticles. Appears dark gray in reflected light. *See also* **Maceral**.

Exothermic reaction: process in which heat is evolved.

Extraneous moisture: surface moisture.

Fine coal: coal pieces less than $\frac{1}{2}$ mm in size. *See also* **Sample**.

Fines: content of fine particles in a coal sample.

Fixed carbon: combustible residue (other than ash) left after the volatile matter is driven off; generally represents the portion of coal that must be burned in the solid state; predominantly carbon but may contain appreciable amounts of sulfur, hydrogen, nitrogen, and oxygen (ASTM D-3172).

Float-and-sink analysis: separation of crushed coal into density fractions using a liquid of predetermined specific gravity; washability curves are prepared from these data (ASTM D-4371).

Fluidity: degree of plasticity exhibited by a sample of coal heated in the absence of air under controlled conditions (ASTM D-1812; ASTM D-2639).

Fluid temperature (ash fluid temperature): temperature at which the coal ash becomes fluid and flows in streams (ASTM D-1857).

Fly ash: airborne particles of ash carried into the atmosphere by stack gases from a power plant.

Fouling: accumulation of small, sticky molten particles of coal ash on a boiler surface.

Free impurities: impurities in coal that exist as individual discrete particles that are not a structural part of the coal and that can be separated from it by coal preparation methods (ASTM D-2234; ASTM D-4915).

Free moisture (surface moisture): coal moisture removed by air drying under standard conditions approximating atmospheric equilibrium; a collection of bubbles and particles on the surface of a pulp in a froth flotation cell (ASTM D-5114).

Free-swelling index: measure of the agglomerating tendency of coal heated to 800°C (1470°F) in a crucible. Coals with a high index are referred to as *coking coals*; those with a low index are referred to as *free-burning coal.*

Friability: tendency of coal particles to break down in size during storage, transportation, or handling; quantitatively expressed as the ratio of average particle size after test to average particle size before test × 100.

Froth flotation: process for cleaning coal fines in which separation from mineral matter is achieved by attachment of the coal to air bubbles in a water medium, allowing the coal to gather in the froth while the mineral matter sinks (ASTM D-5114).

Frother: reagent used in froth flotation to control the size and stability of air bubbles, principally by reducing the surface tension of water (ASTM D-5114).

Fusain: black macroscopic coal constituent (lithotype) that resembles wood charcoal; extremely soft and friable. Also, U.S. Bureau of Mines term for mineral charcoal seen by transmitted light microscopy. *See also* **Coal**.

Fusibility: ability of coal ash to soften and become fluid when heated according to prescribed conditions. *See also* **Ash fusion temperatures**.

Fusinite: microscopic coal constituent (maceral) with well-preserved cell structure and cell cavities empty or occupied by mineral matter. *See also* **Maceral**.

Gravity separation: treatment of coal particles that depends mainly on differences in specific gravity of particles for separation.

Grindability index: number that indicates the ease of pulverizing a coal in comparison to a reference coal. This index is helpful in estimating mill capacity. The two most common methods for determining this index are the Hardgrove grindability method and ball mill grindability method. Coals with a low index are more difficult to pulverize.

Gross calorific value (gross heat of combustion at constant volume): heat produced by combustion of a unit quantity of a solid or liquid fuel when burned at constant volume in an oxygen bomb calorimeter under specified conditions, with the resulting water condensed to a liquid; not applied to gaseous fuels and applies to a volatile liquid fuel only if it is suitably contained during the measurement; closely related to the internal energy of combustion for the same reaction at constant standard temperature and pressure.

Gross sample: *see* **Sample**.

Hard coal: coal with a heat content greater than 10,260 Btu/lb on a moist, ash-free basis. Includes anthracite, bituminous, and the higher-rank subbituminous coals.

Hardgrove grindability index: weight percent of coal retained on a No. 200 sieve after treatment as specified in ASTM D-409.

Heat capacity: energy equivalent.

Heavy media: *see* **Dense media**.

Hemispherical temperature: in the fusibility of coal and coke ash, this is the temperature at which the cone has fused down to a hemispherical lump, at which point the height is one-half the width of the base (ASTM D-1857).

High-temperature tar: heavy distillate from the pyrolysis of coal at a temperature of about 800°C (1470°F).

High-volatile A bituminous coal: *see* **Rank**.

High-volatile B bituminous coal: *see* **Rank**.

High-volatile C bituminous coal: *see* **Rank**.

Hydrogenation: chemical reaction in which hydrogen is added to a substance.

Impure coal: *see* **Coal**.

Increment: small portion of a lot collected by one operation of a sampling device and normally combined with other increments from the lot to make a gross sample (ASTM D-2234; ASTM D-4916).

Inertinite: *see* **Maceral**.

Inertodetrinite: *see* **Maceral**.

Inherent ash: residue remaining from the inherent impurities after ignition (ASTM D-2234).

Inherent impurity: inorganic material in coal that is structurally part of the coal and cannot be separated from it by coal preparation methods (ASTM D-2234).

Inherent moisture: *see* **Moisture**.

Initial deformation temperature: temperature at which the first rounding of the apex of the cone occurs (ASTM D-1857).

ISO: International Organization for Standardization, located in Geneva, Switzerland.

Isoperibol: term used in combustion calorimetry, meaning constant-temperature environment.

Isotropic phase: binder-phase carbon texture that exhibits optical properties (ASTM D-5061).

Jigs: machines that produce stratification of the particles in a bed or particles of differing densities by repeated differential agitation of the bed, the heaviest particles migrating to the lowest layer. The jigging action may be carried out in air or with the bed immersed in water or other liquids.

Laboratory sample: *see* **Sample**.

Light oil: coal tar and coal gas fraction with distillation range between 80 and 210°C (175 to 410°F) containing mainly benzene, with smaller amounts of toluene and xylene.

Lignite: brownish-black woody-structured coal, lower in fixed carbon and higher in volatile matter and oxygen than either anthracite or bituminous coal; similar to the *brown coal* of Europe and Australia. *See also* **Rank**.

Lot: discrete quantity of coal for which the overall quality to a particular precision needs to be determined (ASTM D-2234).

Low-volatile bituminous coal: *see* **Rank**.

Maceral: microscopic petrographic units of coal (ASTM D-2796; ASTM D-2797; ASTM D-2798; ASTM D-2799).

Cutinite: maceral derived from the waxy coatings (cuticles) of leaves and other plant parts.

Exinite: liptinite.

Fusinite: maceral distinguished by the well-preserved original form of plant cell wall structure, intact or broken, with open or mineral-filled cell lumens (cavities).

Inertinite: group of macerals composed of fusinite, inertodetrinite, macrinite, micrinite, sclerotinite, and semifusinite.

Inertodetrinite: maceral occurring as individual, angular, clastic fragments of other inertinite macerals, surrounded by other macerals, commonly vitrinite or minerals, and also distinguished by a reflectance higher than that of associated vitrinite.

Liptinite: group of macerals composed of alginite, cutinite, resinite, and sporinite; derived from secretions of plants and distinguished from one another by morphology.

Macrinite: maceral that is distinguished by a reflectance higher than that of associated vitrinite, absence of recognizable plant cell structure, and a nonangular shape.

Micrinite: maceral that is distinguished by reflectance higher than that of associated vitrinite, absence of recognizable plant cell structure.

Resinite: maceral derived from the resinous secretions and exudates of plant cells.

Sclerotinite: maceral having reflectance between that of fusinite and associated vitrinite and occurring as round or oval cellular bodies or as interlaced tissues derived from fungal remains.

Semifusinite: maceral that is intermediate in reflectance between fusinite and associated vitrinite that shows plant cell wall structure with cavities generally oval or elongated in cross section, but in some specimens less well defined than in fusinite; often, occurs as a transitional material between vitrinite and fusinite.

Sporinite: maceral derived from the waxy coatings (exines) of spores and pollen.

Vitrinite: maceral and maceral group composing all or almost all of the villain and like material occurring in attrital coal as the component of reflectance intermediate between those of exinite and inertinite.

Macrinite: *see* **Maceral**.

Mechanical cell: flotation cell that uses mechanical agitation of a pulp by means of an immersed impeller (rotor) and stator stirring mechanism (ASTM D-5114).

Medium-volatile bituminous coal: *see* **Rank**.

Metaanthracite: *see* **Rank**.

Metallurgical coal: coal that meets the requirements for use in the steelmaking process to manufacture coke; it must be low in ash and sulfur and form coke that is capable of supporting the charge of iron ore and limestone in a blast furnace.

Micrinite: *see* **Maceral**.

Middle (carbolic or creosote) oil: coal tar fraction with a distillation range of 200 to 270°C (390 to 520°F), containing mainly naphthalene, phenol, and cresols.

Middlings: coal of an intermediate specific gravity and quality.

Mineralized coal: *see* **Coal**.

Moisture: total moisture content of a sample customarily determined by adding the moisture loss obtained when air drying the sample and the measured moisture content of the dried sample. Moisture does not represent all of the water present in coal, as water of decomposition (combined water) and hydration are not given off under standardized test conditions.

Inherent moisture: moisture that exists as an integral part of the coal in its natural state, including water in pores but not that present in macroscopically visible fractures (ASTM D-388; ASTM D-1412).

Residual moisture: moisture remaining in the sample after air drying; assayed by determining the mass lost from drying the sample at 104 to 110°C (219 to 230°F) at specified conditions of residence time, atmosphere, particle size, sample mass, and equipment configuration (ASTM D-3173; ASTM D-3302).

Total moisture: all of the moisture in and on a consignment or sample of coal; commonly determined by quantitatively air drying a sample and then assaying residual moisture in the air-dried sample; thus, total moisture is the sum of the air-dry loss and the residual moisture adjusted to an as-received basis (ASTM D-2961; ASTM D-3302).

Natural gas: naturally occurring gas with a heat content over 1000 Btu/ft^3, consisting mainly of methane but also containing smaller amounts of the C_2–C_4 hydrocarbons as well as nitrogen, carbon dioxide, and hydrogen sulfide.

Net calorific value (net heat of combustion at constant pressure): heat produced by combustion of a unit quantity of a solid or liquid fuel when burned, at a constant pressure of 1 atm, under conditions such that all the water in the products remains in the forms of vapor.

Nonbanded coal: *see* **Coal**.

Noncaking coal: coal that does not fuse together or cake when heated but burns freely.

Nonslagging coal: coals whose ash softens at more than 1425°C (2600°F), requiring removal in a solid state; cf. slagging coals whose ash is softened at less than 1205°C (2200°F) and can be discharged in a molten state.

Oil agglomeration: treatment of a suspension of fine coal particles suspended in water with a light hydrocarbon oil so that the particles are preferentially "collected" by the oil, which separates as a floating pasty agglomerate and can be removed by skimming. Developed as a method for recovering fine coal particles by Trent in 1914.

Oxidized coal: coal whose properties have been modified fundamentally as a result of chemisorptions of oxygen in the air or oxygen dissolved in groundwater. Chemisorption is a surface phenomenon rarely detectable by chemical analysis but usually detectable by petrographic examination. It reduces the affinity of coal surfaces for oil and seriously impairs coking, caking, and agglutinating properties.

Peat: partially carbonized plant matter, formed by slow decay in water.

Physical coal cleaning: processes that employ a number of different operations, including crushing, sizing, dewatering and clarifying, and drying, which improve the quality of the fuel by regulating its size and reducing the quantities of ash, sulfur, and other impurities. In this book the term *coal cleaning* is synonymous with the terms *coal preparation, beneficiation*, and *washing*.

Plasticity: property of certain coals when heated in the absence of air. For a relative and a semiquantitative method for determining the relative plastic behavior of coal, refer to ASTM D-2639 and ASTM D-1812, respectively.

Precision: measure of the maximum random error or deviation of a single observation. It may be expressed as the standard error or a multiple thereof, depending on the probability level desired.

Preparation: process of upgrading run-of-mine coal to meet market specifications by washing and sizing.

Prepared coal: any coal, regardless of its top size, that has been cleaned manually or mechanically; includes coal that has been processed over a picking table or air tables through a breaker, jig, or other device which segregates according to size or density (specific gravity) (ASTM D-4749).

Pretreatment: mild oxidation of coal to eliminate caking (agglomeration) tendencies.

Proximate analysis: determination, by prescribed methods, of moisture, volatile matter, fixed carbon (by difference), and ash.

Pulp: fluid mixture of solids and water, also known as slurry (ASTM D-5114).

Random variance of increment collection (unit variance): *see* **Variance**.

Rank: property of coal that is descriptive of degree of coalification (i.e., the stage of metamorphosis of the original vegetal material in the increasing sequence peat, lignite, subbituminous, bituminous, and anthracite) (ASTM D-388).

Anthracite: rank of coal such that on a dry, mineral-matter-free basis, the volatile matter content of the coal is greater than 2% but equal to or less than 8% (or the fixed carbon content is equal to or greater than 92% but less than 98%), and the coal is nonagglomerating.

Anthracitic class: class of rank consisting of semianthracite, anthracite, and meta-anthracite.

Bituminous class: class of rank consisting of high-volatile C bituminous coal, high-volatile B bituminous coal, high-volatile A bituminous coal, medium-volatile bituminous coal, and low-volatile bituminous coal.

High-volatile A bituminous coal: rank of coal such that on a dry, mineral-matter-free basis, the volatile matter content of coal is greater than 31% (or the fixed carbon content is less than 69%), and its gross calorific value is equal to or greater than 14,000 Btu/lb of coal on a moist, mineral-matter-free basis, and the coal is commonly agglomerating.

High-volatile B bituminous coal: rank of coal such that on a moist, mineral-matter-free basis, the gross calorific value of the coal in British thermal units per pound is equal to greater than 13,000 but less than 14,000, and the coal commonly agglomerates.

High-volatile C bituminous coal: rank of coal such that on a moist, mineral-matter-free basis, the gross calorific value of the coal in British thermal units per pound is equal to or greater than 11,500 but less than 13,000, and the coal commonly agglomerates, or equal to or greater than 10,500 but less than 11,500, and the coal agglomerates.

Lignite A: rank of coal such that on a moist, mineral-matter-free basis, the gross calorific value of the coal in British thermal units per pound is equal to or greater than 6300 but less than 8300, and the coal is nonagglomerating.

Lignite B: the rank of coal within the lignite class such that on a moist, mineral-matter-free basis, the gross calorific value of the coal in British thermal units per pound is less than 6300, and the coal is nonagglomerating.

Lignite class: class of rank consisting of lignite A and lignite B.

Low-volatile bituminous coal: rank of coal within the bituminous class such that on a dry, mineral-matter-free basis, the volatile matter content of the coal is greater than 14% but equal to or less than 22% (or the fixed carbon content is equal to or greater than 78% but less than 86%), and the coal commonly agglomerates.

Medium-volatile bituminous coal: rank of coal within the bituminous class such that on a dry, mineral-matter-free basis, the volatile matter content of the coal is greater than 22% but equal to or less than 31% (or the fixed carbon content is equal to or greater than 69% but less than 78%), and the coal commonly agglomerates.

Metaanthracite: rank of coal within the anthracite class such that on a dry, mineral-matter-free basis, the volatile matter content of the coal is equal to or less than 2% (or the fixed carbon is equal to or greater than 98%), and the coal is nonagglomerating.

Semianthracite: rank of coal such that on a dry, mineral-matter-free basis, the volatile matter content of the coal is greater than 8% but equal to or less than 14% (or the fixed carbon content is equal to or greater than 86% but less than 92%), and the coal is nonagglomerating.

Subbituminous class: class of rank consisting of subbituminous A coal, subbituminous B coal, and subbituminous C coal.

Subbituminous A coal: rank of coal such that on a moist, mineral-matter-free basis, the gross calorific value of the coal in British thermal units per pound is equal to or greater than 10,500 but less than 11,500, and the coal is nonagglomerating.

Subbituminous B coal: rank of coal such that on a moist, mineral-matter-free basis, the gross calorific value of the coal in British thermal units per pound is equal to or greater than 9500 but less than 10,500, and the coal is nonagglomerating.

Subbituminous C coal: rank of coal such that on a moist, mineral-matter-free basis, the gross calorific value of the coal in British thermal units per pound is equal to greater than 8300 but less than 9500, and the coal is nonagglomerating.

Raw coal: run-of-mine coal that has been treated by the removal of tramp material, screening, or crushing (ASTM D-4749).

Repeatability: closeness of agreement between test results carried out by one person with one instrument in one laboratory.

Replicate: measurement or observation that is part of a series performed on the same sample.

Reporting:

As-determined basis: analytical data obtained from the analysis sample of coal or coke after conditioning and preparation (ASTM D-2013; ASTM D-3180).

As-received basis: analytical data calculated to the moisture condition of the sample as it arrived at the laboratory, before processing or conditioning (ASTM D-3180).

Dry, ash-free basis: data calculated to a theoretical base of no moisture or ash associated with the sample (ASTM D-3173; ASTM D-3174).

Dry basis: data calculated to a theoretical base of no moisture associated with the sample (ASTM D-3173).

Equilibrium moisture basis: data calculated to the moisture level established as the equilibrium moisture (ASTM D-1412; ASTM D-3180).

Reproducibility: measure of agreement between test results carried out by more than one person with more than one instrument in more than one laboratory.

Run-of-mine coal: coal as it leaves the mine prior to any type of crushing or preparation (ASTM D-4749).

Sample: quantity of material taken from a larger quantity for the purpose of estimating properties or composition of the larger quantity (ASTM D-2234).

Analysis sample: final subsample prepared from the original gross sample (ASTM D-2013; ASTM D-2234).

Coarse coal: portion of a coal sample being subjected to a washability study that is larger than a specific predetermined particle size (ASTM D-4371).

Divided sample: sample that has been reduced in quantity (ASTM D-2013).

Fine coal: portion of a coal sample being subject to a washability study that is smaller than the predetermined particle size (ASTM D-4371).

Gross moisture sample: sample representing one lot of coal and composed of a number of increments on which neither reduction nor division has been performed or a subsample for moisture testing taken.

Gross sample: sample representing one lot of coal and composed of a number of increments on which neither reduction nor division has been performed (ASTM D-2013; ASTM D-D 2234).

Laboratory sample: sample, not less than the permissible weight delivered to the laboratory for further preparation and analysis (ASTM D-2013).

Representative sample: sample collected in such a manner that every particle in the lot to be sampled is equally represented in the gross or divided sample (ASTM D-2013; ASTM D-2234; ASTM D-4916).

Subsample: sample taken from another sample (ASTM D-2013; ASTM D-2234; ASTM D-4916).

Unbiased sample (representative sample): sample free of bias (ASTM D-2013; ASTM D-2234; ASTM D-4916).

Sample division: process whereby a sample is reduced in weight without change in particle size (ASTM D-2013; ASTM D-2234; ASTM D-4916).

Sample reduction: process whereby a sample is reduced in particle size by crushing or grinding without significant change (ASTM D-2013).

Sample preparation: process that may include air drying, crushing, division, and mixing of a gross sample for the purpose of obtaining an unbiased analysis sample (ASTM D-2013).

Sclerotinite: *see* **Maceral**.

Seam: underground layer of coal or other mineral of any thickness.

Seam moisture: inherent moisture.

Semianthracite: *see* **Rank**.

Semifusinite: *see* **Maceral**.

Significant loss: any loss that introduces a bias in final results that is of appreciable economic importance to the parties concerned (ASTM D-2013; ASTM D-2234; ASTM D-4915; ASTM D-4916).

Size consist: particle size distribution of coal (ASTM D-2013; ASTM D-2234; ASTM D-4915; ASTM D-4916).

Slag: coal ash that is or has been in a molten state.

Slagging: accumulation of coal ash on the wall tubes of a coal-fired boiler furnace, forming a solid layer of ash residue and interfering with heat transfer.

Slagging coal: coal whose ash is softened at less than 1205°C (2200°F) and can be discharged in a molten state; cf. nonslagging coal, whose ash softens at more than 1425°F (2600°F), requiring removal in a solid state.

Slurry: mixture of pulverized insoluble material and water.

Slurry pipeline: pipeline that can transport a coal–water mixture for long distances.

Softening temperature: temperature at which the cone has fused down to a spherical lump in which the height is equal to the width at the base (ASTM D-1857).

Solids concentration: ratio of the weight (mass) of solids to the sum of the weight of solids plus water (ASTM D-5114).

Spacing of increments: intervals between increments; systematic and random. With systematic spacing being preferable.

Systematic spacing: spacing in which the movements of individual increment collection are spaced evenly in time or in position over the lot.

Random spacing: spacing in which the increments are spaced at random in time or in position over the lot (ASTM D-4916).

Sparging: bubbling a gas into the bottom of a pool of liquid.

Sparking fuels: fuels that do not yield a coherent cake as residue in the volatile matter determination but evolve gaseous products at a rate sufficient to carry solid particles mechanically out of the crucible when heated at the standard rate; usually, all low-rank noncaking coal and lignite; may also include those anthracite, semianthracite, and bituminous coals that lose solid particles as described above; particles escaping at the higher temperatures may become incandescent and spark as they are emitted (ASTM D-3175).

Specific energy: energy per unit of throughput required to reduce feed material to a desired product size.

Specific gravity: ratio of weight per unit volume of a substance to the weight of the same unit volume of water.

Splint coal: U.S. Bureau of Mines term for durain with some vitrain and clarain and a small amount of fusain.

Spontaneous combustion: self-ignition of coal through oxidation under very specific conditions; different types of coal vary in their tendency toward self-ignition.

Sporinite: *see* **Maceral**.

Standard deviation: usual measure of the dispersion of observed values or results expressed as the positive square root of the variance (ASTM D-2013; ASTM D-2234).

Steam coal: coal used in electrical power plants to produce the steam that runs generators.

Subbituminous coal: glossy-black-weathering and nonagglomerating coal that is lower in fixed carbon than bituminous coal, with more volatile matter and oxygen. *See also* **Coal**.

Subsample: *see* **Sample**.

Sulfur (total sulfur): sulfur found in coal as iron pyrites, sulfates, and organic compounds. It is undesirable because the sulfur oxides formed when it burns contribute to air pollution, and sulfur compounds contribute to combustion-system corrosion and deposits.

Sulfur forms: analytical percentage by weight of coal sulfate, pyrite, and organic sulfur.

Tailings: underflow product from coal froth flotation (ASTM D-5114).

Test portion: quantity of coal that is representative of the analysis sample and sufficient to obtain a single test result for the property or properties to be measured.

Test specimen: test portion.

Test unit: analysis sample.

Top size: opening of the smallest screen in the series on which is retained less than 5% of the sample (ASTM D-2013; ASTM D-2234; ASTM D-4749; ASTM D-4916); should not be confused with the size of the largest particle in the lot (ASTM D-4749).

Total carbon: carbon content remaining in the solid products derived from the combustion or reaction of coal, coal by-products, or coke, inclusive of carbonate in any form (ASTM D-6316).

Total moisture: *see* **Moisture**.

Total variance: *see* **Variance**.

Trace element: any element present in minute quantities, such as lead and mercury.

Ultimate analysis: analytical percentage by weight of coal carbon, hydrogen, nitrogen, sulfur, oxygen, and ash. *In the case of coal and coke*, determination of carbon and hydrogen in the material, as found in the gaseous products of its complete combustion, the determinations of sulfur, nitrogen, and ash in the material as a whole, and the calculation of oxygen by difference.

Unbiased sample: *see* **Sample**.

Variance: mean square of deviations (or errors) of a set of observations; the sum of squared deviations (or errors) of individual observations with respect to their arithmetic mean divided by the number of observations less one (degrees of freedom); the square of the standard deviation (or standard error) (ASTM D-2013; ASTM D-2234).

Random variance of increment collection (unit variance): theoretical variance calculated for a uniformly mixed lot and extrapolated to increment size (ASTM D-2234).

Segregation variance of increment collection: variance caused by nonrandom distribution of ash content or other constituent in the lot (ASTM D-2234).

Total variance: overall variance resulting from collecting single increments and including division and analysis of the single increments (ASTM D-2013; ASTM D-2234).

Variance of analysis: variance caused by chance errors (deviations) of analysis (ASTM D-2013).

Variance of division: variance caused by chance errors (deviations) of sample division (ASTM D-2013).

Variance of division and analysis: variance caused by the combined chance errors of division and analysis (ASTM D-2013).

Vitrain: macroscopic coal constituent (lithotype) that appear as brilliant black bands of uniform appearance. *See also* **Coal**.

Vitrinite: microscopic coal constituent (maceral) that appears translucent by transmitted light and gray in reflected light; termed *anthraxylon* when viewed by transmitted light. *See also* **Maceral**.

Vitrinite type: reflectance classes of vitrinite that span 0.1% reflectance intervals (ASTM D-5061).

Volatile matter: hydrogen, carbon monoxide, methane, tar, other hydrocarbons, carbon dioxide, and water obtained on thermal decomposition of coal under prescribed conditions (ASTM D-3175).

Washability analysis: procedure used in a laboratory before preparation plant design to determine the cleaning processes to be employed and used during normal operation to evaluate the performance of the cleaning equipment and the amenability of the raw coal feed to the cleaning processes chosen (ASTM D-4371).

Water equivalent: energy equivalent.

Weathering: action of air and water on coal in surface stockpiles, causing size reduction, oxidation, and decreases of any caking or coking properties.

Wet sieving: method of sieving coal that uses water as a medium for facilitating segregation of a sample into particle size (ASTM D-4749).

Yield: weight percent of the feed that reports to the concentrate (ASTM D-5114).

Common Conversion Factors

1 atmosphere = 760 mmHg = 14.696 psi = 29.91 inches Hg = 1.0133 bar = 33.899 feet of water

1 barrel (oil) = 42 gallons = 5.6146 cubic feet

1 Btu = 778.26 ft-lb = 1 Btu = 1054.35 joules = $1.055055853 \times 10^{10}$ ergs

1 calorie per gram = 4.184 joules

1 centimeter = 0.39 inch

1 centipoise × 2.42 = lb mass/ft-hr, viscosity

1 centipoise × 0.000672 = lb mass/ft-sec, viscosity

1 cubic foot = 28,317 cubic centimeters = 7.4805 gallons

1 erg = 2.39006×10^{-8} calories

1 gallon (U.S.) = 231 cubic inches = 3.78541 liters

1 gallon (Imperial) = 277.419 cubic inches = 4.45609 liters

1 gram = 1 kilogram $\times 10^{-3}$ = 0.0022046 pound

1 inch = 2.54 cm = 24.4 mm

1 joule = 10^{7} ergs = 0.239006 calorie

1 kilogram = 2.20462 pounds

1 liter = 0.26417 gallon = 61.0237 cubic inches

1 meter = 100 cm = 1000 mm = 10^{6} micrometers = 10^{10} angstroms (Å) = 3.2008 feet

1 micrometer = 10^{-4} cm

1 ounce = 28.35 grams

1 pascal = 9.86923×10^{-6} atmospheres = 1.45×10^{-4} psi

1 pound = 453.592 grams = 0.453592 kilogram = 7000 grains

1 ton (long) = 2240 pounds

1 ton (metric) = 2205 pounds = 1000 kilograms

1 ton (short or net) = 2000 pounds = 907.185 kilograms = 0.970 metric ton

Handbook of Coal Analysis, by James G. Speight
ISBN 0-471-52273-2

Units of ten:

- 10^{-12} pico
- 10^{-9} nano
- 10^{-6} mico
- 10^{-3} milli
- 10^{3} kilo
- 10^{6} mega
- 10^{9} giga
- 10^{12} tera

1 yard = 0.9144 meter

INDEX

Handbook of Coal Analysis, by James G. Speight
ISBN 0-471-52273-2

CHEMICAL ANALYSIS

A SERIES OF MONOGRAPHS ON ANALYTICAL CHEMISTRY
AND ITS APPLICATIONS

Series Editor
J. D. WINEFORDNER

Vol. 1 **The Analytical Chemistry of Industrial Poisons, Hazards, and Solvents**. *Second Edition.* By the late Morris B. Jacobs
Vol. 2 **Chromatographic Adsorption Analysis**. By Harold H. Strain (*out of print*)
Vol. 3 **Photometric Determination of Traces of Metals**. *Fourth Edition*
Part I: General Aspects. By E. B. Sandell and Hiroshi Onishi
Part IIA: Individual Metals, Aluminum to Lithium. By Hiroshi Onishi
Part IIB: Individual Metals, Magnesium to Zirconium. By Hiroshi Onishi
Vol. 4 **Organic Reagents Used in Gravimetric and Volumetric Analysis**. By John F. Flagg (*out of print*)
Vol. 5 **Aquametry: A Treatise on Methods for the Determination of Water**. *Second Edition* (*in three parts*). By John Mitchell, Jr. and Donald Milton Smith
Vol. 6 **Analysis of Insecticides and Acaricides**. By Francis A. Gunther and Roger C. Blinn (*out of print*)
Vol. 7 **Chemical Analysis of Industrial Solvents**. By the late Morris B. Jacobs and Leopold Schetlan
Vol. 8 **Colorimetric Determination of Nonmetals**. *Second Edition.* Edited by the late David F. Boltz and James A. Howell
Vol. 9 **Analytical Chemistry of Titanium Metals and Compounds**. By Maurice Codell
Vol. 10 **The Chemical Analysis of Air Pollutants**. By the late Morris B. Jacobs
Vol. 11 **X-Ray Spectrochemical Analysis**. *Second Edition.* By L. S. Birks
Vol. 12 **Systematic Analysis of Surface-Active Agents**. *Second Edition.* By Milton J. Rosen and Henry A. Goldsmith
Vol. 13 **Alternating Current Polarography and Tensammetry**. By B. Breyer and H.H.Bauer
Vol. 14 **Flame Photometry**. By R. Herrmann and J. Alkemade
Vol. 15 **The Titration of Organic Compounds** (*in two parts*). By M. R. F. Ashworth
Vol. 16 **Complexation in Analytical Chemistry: A Guide for the Critical Selection of Analytical Methods Based on Complexation Reactions**. By the late Anders Ringbom
Vol. 17 **Electron Probe Microanalysis**. *Second Edition.* By L. S. Birks
Vol. 18 **Organic Complexing Reagents: Structure, Behavior, and Application to Inorganic Analysis**. By D. D. Perrin
Vol. 19 **Thermal Analysis**. *Third Edition*. By Wesley Wm.Wendlandt
Vol. 20 **Amperometric Titrations**. By John T. Stock
Vol. 21 **Reflctance Spectroscopy**. By Wesley Wm.Wendlandt and Harry G. Hecht
Vol. 22 **The Analytical Toxicology of Industrial Inorganic Poisons**. By the late Morris B. Jacobs
Vol. 23 **The Formation and Properties of Precipitates**. By Alan G.Walton
Vol. 24 **Kinetics in Analytical Chemistry**. By Harry B. Mark, Jr. and Garry A. Rechnitz
Vol. 25 **Atomic Absorption Spectroscopy**. *Second Edition.* By Morris Slavin
Vol. 26 **Characterization of Organometallic Compounds** (*in two parts*). Edited by Minoru Tsutsui
Vol. 27 **Rock and Mineral Analysis**. *Second Edition.* By Wesley M. Johnson and John A. Maxwell
Vol. 28 **The Analytical Chemistry of Nitrogen and Its Compounds** (*in two parts*). Edited by C. A. Streuli and Philip R.Averell
Vol. 29 **The Analytical Chemistry of Sulfur and Its Compounds** (*in three parts*). By J. H. Karchmer
Vol. 30 **Ultramicro Elemental Analysis**. By Güther Toölg
Vol. 31 **Photometric Organic Analysis** (*in two parts*). By Eugene Sawicki
Vol. 32 **Determination of Organic Compounds: Methods and Procedures**. By Frederick T. Weiss
Vol. 33 **Masking and Demasking of Chemical Reactions**. By D. D. Perrin

Vol. 34. **Neutron Activation Analysis**. By D. De Soete, R. Gijbels, and J. Hoste
Vol. 35 **Laser Raman Spectroscopy**. By Marvin C. Tobin
Vol. 36 **Emission Spectrochemical Analysis**. By Morris Slavin
Vol. 37 **Analytical Chemistry of Phosphorus Compounds**. Edited by M. Halmann
Vol. 38 **Luminescence Spectrometry in Analytical Chemistry**. By J. D.Winefordner, S. G. Schulman, and T. C. O'Haver
Vol. 39. **Activation Analysis with Neutron Generators**. By Sam S. Nargolwalla and Edwin P. Przybylowicz
Vol. 40 **Determination of Gaseous Elements in Metals**. Edited by Lynn L. Lewis, Laben M. Melnick, and Ben D. Holt
Vol. 41 **Analysis of Silicones**. Edited by A. Lee Smith
Vol. 42 **Foundations of Ultracentrifugal Analysis**. By H. Fujita
Vol. 43 **Chemical Infrared Fourier Transform Spectroscopy**. By Peter R. Griffiths
Vol. 44 **Microscale Manipulations in Chemistry**. By T. S. Ma and V. Horak
Vol. 45 **Thermometric Titrations**. By J. Barthel
Vol. 46 **Trace Analysis: Spectroscopic Methods for Elements**. Edited by J. D.Winefordner
Vol. 47 **Contamination Control in Trace Element Analysis**. By Morris Zief and James W. Mitchell
Vol. 48 **Analytical Applications of NMR**. By D. E. Leyden and R. H. Cox
Vol. 49 **Measurement of Dissolved Oxygen**. By Michael L. Hitchman
Vol. 50 **Analytical Laser Spectroscopy**. Edited by Nicolo Omenetto
Vol. 51 **Trace Element Analysis of Geological Materials**. By Roger D. Reeves and Robert R. Brooks
Vol. 52 **Chemical Analysis by Microwave Rotational Spectroscopy**. By Ravi Varma and Lawrence W. Hrubesh
Vol. 53 **Information Theory as Applied to Chemical Analysis**. By Karl Eckschlager and Vladimir Stepanek
Vol. 54 **Applied Infrared Spectroscopy: Fundamentals, Techniques, and Analytical Problemsolving**. By A. Lee Smith
Vol. 55 **Archaeological Chemistry**. By Zvi Goffer
Vol. 56 **Immobilized Enzymes in Analytical and Clinical Chemistry**. By P. W. Carr and L. D. Bowers
Vol. 57 **Photoacoustics and Photoacoustic Spectroscopy**. By Allan Rosencwaig
Vol. 58 **Analysis of Pesticide Residues**. Edited by H. Anson Moye
Vol. 59 **Affity Chromatography**. By William H. Scouten
Vol. 60 **Quality Control in Analytical Chemistry**. *Second Edition.* By G. Kateman and L. Buydens
Vol. 61 **Direct Characterization of Fineparticles**. By Brian H. Kaye
Vol. 62 **Flow Injection Analysis**. By J. Ruzicka and E. H. Hansen
Vol. 63 **Applied Electron Spectroscopy for Chemical Analysis**. Edited by Hassan Windawi and Floyd Ho
Vol. 64 **Analytical Aspects of Environmental Chemistry**. Edited by David F. S. Natusch and Philip K. Hopke
Vol. 65 **The Interpretation of Analytical Chemical Data by the Use of Cluster Analysis**. By D. Luc Massart and Leonard Kaufman
Vol. 66 **Solid Phase Biochemistry: Analytical and Synthetic Aspects**. Edited by William H. Scouten
Vol. 67 **An Introduction to Photoelectron Spectroscopy**. By Pradip K. Ghosh
Vol. 68 **Room Temperature Phosphorimetry for Chemical Analysis**. By Tuan Vo-Dinh
Vol. 69 **Potentiometry and Potentiometric Titrations**. By E. P. Serjeant
Vol. 70 **Design and Application of Process Analyzer Systems**. By Paul E. Mix
Vol. 71 **Analysis of Organic and Biological Surfaces**. Edited by Patrick Echlin
Vol. 72 **Small Bore Liquid Chromatography Columns: Their Properties and Uses**. Edited by Raymond P.W. Scott
Vol. 73 **Modern Methods of Particle Size Analysis**. Edited by Howard G. Barth
Vol. 74 **Auger Electron Spectroscopy**. By Michael Thompson, M. D. Baker, Alec Christie, and J. F. Tyson
Vol. 75 **Spot Test Analysis: Clinical, Environmental, Forensic and Geochemical Applications**. By Ervin Jungreis
Vol. 76 **Receptor Modeling in Environmental Chemistry**. By Philip K. Hopke

Vol. 77 **Molecular Luminescence Spectroscopy: Methods and Applications** (*in three parts*). Edited by Stephen G. Schulman
Vol. 78 **Inorganic Chromatographic Analysis**. Edited by John C. MacDonald
Vol. 79 **Analytical Solution Calorimetry**. Edited by J. K. Grime
Vol. 80 **Selected Methods of Trace Metal Analysis: Biological and Environmental Samples**. By Jon C.VanLoon
Vol. 81 **The Analysis of Extraterrestrial Materials**. By Isidore Adler
Vol. 82 **Chemometrics**. By Muhammad A. Sharaf, Deborah L. Illman, and Bruce R. Kowalski
Vol. 83 **Fourier Transform Infrared Spectrometry**. By Peter R. Griffiths and James A. de Haseth
Vol. 84 **Trace Analysis: Spectroscopic Methods for Molecules**. Edited by Gary Christian and James B. Callis
Vol. 85 **Ultratrace Analysis of Pharmaceuticals and Other Compounds of Interest**. Edited by S. Ahuja
Vol. 86 **Secondary Ion Mass Spectrometry: Basic Concepts, Instrumental Aspects, Applications and Trends**. By A. Benninghoven, F. G. Rüenauer, and H.W.Werner
Vol. 87 **Analytical Applications of Lasers**. Edited by Edward H. Piepmeier
Vol. 88 **Applied Geochemical Analysis**. By C. O. Ingamells and F. F. Pitard
Vol. 89 **Detectors for Liquid Chromatography**. Edited by Edward S.Yeung
Vol. 90 **Inductively Coupled Plasma Emission Spectroscopy: Part 1: Methodology, Instrumentation, and Performance; Part II: Applications and Fundamentals**. Edited by J. M. Boumans
Vol. 91 **Applications of New Mass Spectrometry Techniques in Pesticide Chemistry**. Edited by Joseph Rosen
Vol. 92 **X-Ray Absorption: Principles, Applications, Techniques of EXAFS, SEXAFS, and XANES**. Edited by D. C. Konnigsberger
Vol. 93 **Quantitative Structure-Chromatographic Retention Relationships**. By Roman Kaliszan
Vol. 94 **Laser Remote Chemical Analysis**. Edited by Raymond M. Measures
Vol. 95 **Inorganic Mass Spectrometry**. Edited by F.Adams, R.Gijbels, and R.Van Grieken
Vol. 96 **Kinetic Aspects of Analytical Chemistry**. By Horacio A. Mottola
Vol. 97 **Two-Dimensional NMR Spectroscopy**. By Jan Schraml and Jon M. Bellama
Vol. 98 **High Performance Liquid Chromatography**. Edited by Phyllis R. Brown and Richard A. Hartwick
Vol. 99 **X-Ray Fluorescence Spectrometry**. By Ron Jenkins
Vol. 100 **Analytical Aspects of Drug Testing**. Edited by Dale G. Deustch
Vol. 101 **Chemical Analysis of Polycyclic Aromatic Compounds**. Edited by Tuan Vo-Dinh
Vol. 102 **Quadrupole Storage Mass Spectrometry**. By Raymond E. March and Richard J. Hughes (*out of print: see Vol. 165)*
Vol. 103 **Determination of Molecular Weight**. Edited by Anthony R. Cooper
Vol. 104 **Selectivity and Detectability Optimization in HPLC**. By Satinder Ahuja
Vol. 105 **Laser Microanalysis**. By Lieselotte Moenke-Blankenburg
Vol. 106 **Clinical Chemistry**. Edited by E. Howard Taylor
Vol. 107 **Multielement Detection Systems for Spectrochemical Analysis**. By Kenneth W. Busch and Marianna A. Busch
Vol. 108 **Planar Chromatography in the Life Sciences**. Edited by Joseph C. Touchstone
Vol. 109 **Fluorometric Analysis in Biomedical Chemistry: Trends and Techniques Including HPLC Applications**. By Norio Ichinose, George Schwedt, Frank Michael Schnepel, and Kyoko Adochi
Vol. 110 **An Introduction to Laboratory Automation**. By Victor Cerdá and Guillermo Ramis
Vol. 111 **Gas Chromatography: Biochemical, Biomedical, and Clinical Applications**. Edited by Ray E. Clement
Vol. 112 **The Analytical Chemistry of Silicones**. Edited by A. Lee Smith
Vol. 113 **Modern Methods of Polymer Characterization**. Edited by Howard G. Barth and Jimmy W. Mays
Vol. 114 **Analytical Raman Spectroscopy**. Edited by Jeanette Graselli and Bernard J. Bulkin
Vol. 115 **Trace and Ultratrace Analysis by HPLC**. By Satinder Ahuja
Vol. 116 **Radiochemistry and Nuclear Methods of Analysis**. By William D. Ehmann and Diane E.Vance

Vol. 117 **Applications of Fluorescence in Immunoassays**. By Ilkka Hemmila
Vol. 118 **Principles and Practice of Spectroscopic Calibration**. By Howard Mark
Vol. 119 **Activation Spectrometry in Chemical Analysis**. By S. J. Parry
Vol. 120 **Remote Sensing by Fourier Transform Spectrometry**. By Reinhard Beer
Vol. 121 **Detectors for Capillary Chromatography**. Edited by Herbert H. Hill and Dennis McMinn
Vol. 122 **Photochemical Vapor Deposition**. By J. G. Eden
Vol. 123 **Statistical Methods in Analytical Chemistry**. By Peter C. Meier and Richard Züd
Vol. 124 **Laser Ionization Mass Analysis**. Edited by Akos Vertes, Renaat Gijbels, and Fred Adams
Vol. 125 **Physics and Chemistry of Solid State Sensor Devices**. By Andreas Mandelis and Constantinos Christofides
Vol. 126 **Electroanalytical Stripping Methods**. By Khjena Z. Brainina and E. Neyman
Vol. 127 **Air Monitoring by Spectroscopic Techniques**. Edited by Markus W. Sigrist
Vol. 128 **Information Theory in Analytical Chemistry**. By Karel Eckschlager and Klaus Danzer
Vol. 129 **Flame Chemiluminescence Analysis by Molecular Emission Cavity Detection**. Edited by David Stiles, Anthony Calokerinos, and Alan Townshend
Vol. 130 **Hydride Generation Atomic Absorption Spectrometry**. Edited by Jiri Dedina and Dimiter L. Tsalev
Vol. 131 **Selective Detectors: Environmental, Industrial, and Biomedical Applications**. Edited by Robert E. Sievers
Vol. 132 **High-Speed Countercurrent Chromatography**. Edited by Yoichiro Ito and Walter D. Conway
Vol. 133 **Particle-Induced X-Ray Emission Spectrometry**. By Sven A. E. Johansson, John L. Campbell, and Klas G. Malmqvist
Vol. 134 **Photothermal Spectroscopy Methods for Chemical Analysis**. By Stephen E. Bialkowski
Vol. 135 **Element Speciation in Bioinorganic Chemistry**. Edited by Sergio Caroli
Vol. 136 **Laser-Enhanced Ionization Spectrometry**. Edited by John C. Travis and Gregory C. Turk
Vol. 137 **Fluorescence Imaging Spectroscopy and Microscopy**. Edited by Xue Feng Wang and Brian Herman
Vol. 138 **Introduction to X-Ray Powder Diffractometry**. By Ron Jenkins and Robert L. Snyder
Vol. 139 **Modern Techniques in Electroanalysis**. Edited by Petr Vanýek
Vol. 140 **Total-Reflction X-Ray Fluorescence Analysis**. By Reinhold Klockenkamper
Vol. 141 **Spot Test Analysis: Clinical, Environmental, Forensic, and Geochemical Applications**. *Second Edition*. By Ervin Jungreis
Vol. 142 **The Impact of Stereochemistry on Drug Development and Use**. Edited by Hassan Y. Aboul-Enein and Irving W.Wainer
Vol. 143 **Macrocyclic Compounds in Analytical Chemistry**. Edited by Yury A. Zolotov
Vol. 144 **Surface-Launched Acoustic Wave Sensors: Chemical Sensing and Thin-Film Characterization**. By Michael Thompson and David Stone
Vol. 145 **Modern Isotope Ratio Mass Spectrometry**. Edited by T. J. Platzner
Vol. 146 **High Performance Capillary Electrophoresis: Theory, Techniques, and Applications**. Edited by Morteza G. Khaledi
Vol. 147 **Solid Phase Extraction: Principles and Practice**. By E. M. Thurman
Vol. 148 **Commercial Biosensors: Applications to Clinical, Bioprocess and Environmental Samples**. Edited by Graham Ramsay
Vol. 149 **A Practical Guide to Graphite Furnace Atomic Absorption Spectrometry**. By David J. Butcher and Joseph Sneddon
Vol. 150 **Principles of Chemical and Biological Sensors**. Edited by Dermot Diamond
Vol. 151 **Pesticide Residue in Foods: Methods, Technologies, and Regulations**. By W. George Fong, H. Anson Moye, James N. Seiber, and John P. Toth
Vol. 152 **X-Ray Fluorescence Spectrometry**. *Second Edition*. By Ron Jenkins
Vol. 153 **Statistical Methods in Analytical Chemistry**. *Second Edition*. By Peter C. Meier and Richard E. Züd
Vol. 154 **Modern Analytical Methodologies in Fat- and Water-Soluble Vitamins**. Edited by Won O. Song, Gary R. Beecher, and Ronald R. Eitenmiller
Vol. 155 **Modern Analytical Methods in Art and Archaeology**. Edited by Enrico Ciliberto and Guiseppe Spoto

Vol. 156 **Shpol'skii Spectroscopy and Other Site Selection Methods: Applications in Environmental Analysis, Bioanalytical Chemistry and Chemical Physics**. Edited by C. Gooijer, F. Ariese and J.W. Hofstraat

Vol. 157 **Raman Spectroscopy for Chemical Analysis**. By Richard L. McCreery

Vol. 158 **Large (C>=24) Polycyclic Aromatic Hydrocarbons: Chemistry and Analysis**. By John C. Fetzer

Vol. 159 **Handbook of Petroleum Analysis**. By James G. Speight

Vol. 160 **Handbook of Petroleum Product Analysis**. By James G. Speight

Vol. 161 **Photoacoustic Infrared Spectroscopy**. By Kirk H. Michaelian

Vol. 162 **Sample Preparation Techniques in Analytical Chemistry**. Edited by Somenath Mitra

Vol. 163 **Analysis and Purification Methods in Combination Chemistry**. Edited by Bing Yan

Vol. 164 **Chemometrics: From Basics to Wavelet Transform**. By Foo-tim Chau, Yi-Zeng Liang, Junbin Gao, and Xue-guang Shao

Vol. 165 **Quadrupole Ion Trap Mass Spectrometry**. *Second Edition.* By Raymond E. March and John F. J. Todd

Vol 166 **Handbook of Coal Analysis**. By James G. Speight